Resource Sharing in Biomedical Research

Committee on
Resource Sharing in Biomedical Research

Division of Health Sciences Policy

INSTITUTE OF MEDICINE

Kenneth I. Berns, Enriqueta C. Bond,
and Frederick J. Manning, *Editors*

NATIONAL ACADEMY PRESS
Washington, D.C. 1996

NATIONAL ACADEMY PRESS • 2101 Constitution Avenue, NW • Washington, DC 20418

NOTICE: The project that is the subject of this report was approved by the Governing Board of the National Research Council, whose members are drawn from the councils of the National Academy of Sciences, the National Academy of Engineering, and the Institute of Medicine. The members of the committee responsible for the report were chosen for their special competences and with regard for appropriate balance.

This report has been reviewed by a group other than the authors according to procedures approved by a Report Review Committee consisting of members of the National Academy of Sciences, the National Academy of Engineering, and the Institute of Medicine.

The Institute of Medicine was chartered in 1970 by the National Academy of Sciences to enlist distinguished members of the appropriate professions in the examination of policy matters pertaining to the health of the public. In this, the Institute acts under both the Academy's 1863 congressional charter responsibility to be an adviser to the federal government and its own initiative in identifying issues of medical care, research, and education. Dr. Kenneth I. Shine is president of the Institute of Medicine.

Support for this project was provided by the National Research Council.

Library of Congress Catalog Card No. 96-71100
International Standard Book Number 0-309-05582-2

Additional copies of this report are available from: National Academy Press, Lock Box 285, 2101 Constitution Avenue, N.W., Washington, DC 20055.

Call (800) 624-6242 or (202) 334-3313 (in the Washington metropolitan area), or visit the NAP on-line bookstore at **http://www.nap.edu/nap/bookstore**.
Call (202) 334-2352 for more information on the other activities of the Institute of Medicine, or visit the IOM home page at **http://www.nas.edu/iom**.

Copyright 1996 by the National Academy of Sciences. All rights reserved.
Printed in the United States of America.

The serpent has been a symbol of long life, healing, and knowledge among almost all cultures and religions since the beginning of recorded history. The image adopted as a logotype by the Institute of Medicine is based on a relief carving from ancient Greece, now held by the Staatlichemuseen in Berlin.

Cover design by Francesca Moghari.

COMMITTEE ON RESOURCE SHARING IN BIOMEDICAL RESEARCH

KENNETH I. BERNS, *Cochair*,*† Professor and Chairman, Department of Microbiology, Cornell University Medical College

ENRIQUETA C. BOND, *Cochair*, President, The Burroughs Wellcome Fund, Morrisville, North Carolina

CHARLES R. CANTOR, Director, Center for Advanced Biotechnology and Professor of Biomedical Engineering and Biophysics, Boston University

LINDA C. CORK,* Chairman, Department of Comparative Medicine, Stanford University School of Medicine

DAVID W. MARTIN, JR., President, Lynx Therapeutics, Inc., Hayward, California

FRANCIS J. MEYER, Associate Vice-Chancellor and Director, Office of Technology Development, University of North Carolina at Chapel Hill

RUSSELL ROSS,* Professor, Department of Pathology and Adjunct Professor, Department of Biochemistry, Center for Vascular Biology, University of Washington School of Medicine, Seattle

MICHAEL SAAG, Associate Professor, Department of Medicine and Associate Director, General Clinical Research Center, University of Alabama at Birmingham

Study Staff

VALERIE SETLOW, Director, Division of Health Science Policy, Institute of Medicine

ERIC FISCHER, Director, Board on Biology, Commission on Life Sciences, National Academy of Sciences

FREDERICK J. MANNING, Study Director

MARGO CULLEN, Project Assistant

JAMAINE TINKER, Financial Associate

---

* Institute of Medicine member.
† National Academy of Sciences member.

Preface

The United States is entering an era when, more than ever, the sharing of resources and information may be critical to scientific progress. Research and development programs are unlikely to escape continuing efforts to control federal spending, making optimal use of resources an imperative. Although competition undoubtedly has been a vital factor in the continuing excellence of American science, every dollar saved by avoiding duplication and producing economies of scale will become increasingly important as funding declines. As an initial step in exploring how best to do that, the National Research Council provided support for a workshop aimed at developing consensus within the research community about critical issues related to the sharing of resources. An eight-member steering committee selected to provide a broad spectrum of experience with shared resources was charged with planning the workshop and providing this report on the workshop and the committee's conclusions and recommendations. The specific goals of the workshop, which was held in Washington, D.C., January 22–23, 1996, were to examine the status of resource sharing in one or more categories of "resource" in the biomedical sciences, to identify critical barriers and opportunities, and to develop a consensus about what needs to be done in this arena.

Although the committee bears full responsibility for the conclusions and recommendations in the report, we would be seriously remiss if we failed to acknowledge the contributions of numerous others to both the planning and the conduct of the workshop. First, we owe thanks to the small group of "liaisons" from federal agencies and professional societies who helped us decide on the form and content of the workshop: James Brown of the National Science Foundation, Maxwell Cowan of the Howard Hughes Medical Institute, Mark Frankel of the American Association for the Advancement of Science, Allan Shipp of the Association of American Medical Colleges, Marvin Snyder of the

Federation of American Societies for Experimental Biology, David Thomasson of the Department of Energy, Fred Tyner of the Army Medical Research and Material Command, and Judith Vaitukaitis of the National Institutes of Health (NIH).

A second group whose hard work made the workshop a success is comprised of the featured speakers—the "presenters" from whom we learned so much about the rewards and trials of resource sharing in today's biomedical research environment: Bruce Alberts of the National Academy of Sciences, Raymond Cypess of the American Type Culture Collection, Chris Somerville of the Carnegie Institution and the Multinational Coordinated *Arabidopsis Thaliana* Genome Research Project, Muriel Davisson of The Jackson Laboratory, William Morton of the Washington Regional Primate Research Center, Steven Ealick of the Macromolecular Crystallography Resource at Cornell, Anthony Carrano of Lawrence Livermore National Laboratory, David Barry of Triangle Pharmaceuticals, Eugene Sokourenko of LabSearch International, Herbert Tabor of NIH and the *Journal of Biological Chemistry*, Jerome Kassirer of the *New England Journal of Medicine*, and Harold Varmus of NIH.

Third, as committee cochairs we are acutely aware of the contributions that the Institute of Medicine staff have made to the success of the study. Special thanks are due to Project Assistant Margo Cullen, who made our travel and meetings as comfortable and convenient as possible and provided outstanding administrative support both at the meetings and in the painstaking production of the final report. We are particularly grateful to Study Director Rick Manning for his skilled and professional support in shepherding the committee through its task.

Finally, we would like to acknowledge the individual and collective efforts of the committee members. It was a pleasure to have worked with this group of busy but unselfish professionals who volunteered their valuable time in sharing their knowledge and experience with their fellow scientists.

Kenneth I. Berns, *Cochair*
Enriqueta C. Bond, *Cochair*

Contents

EXECUTIVE SUMMARY 1

1   INTRODUCTION 13
    Competition for Funds, 14
    Incentives for Scientists, 14
    Nationalism, 16
    Methods and Goals of This Study, 17
    References, 20

2   THE AMERICAN TYPE CULTURE COLLECTION 23
    General Facilities, 24
    Programs, 25
    Ownership and Access Issues, 29
    Cost Issues, 30
    Other Issues and Problems, 32

3   THE MULTINATIONAL COORDINATED
    ARABIDOPSIS THALIANA GENOME
    RESEARCH PROJECT 33
    Project Elements, 33
    Ownership and Access Issues, 35
    Cost Issues, 37
    Other Issues and Problems, 37

4   THE JACKSON LABORATORY 39
    Animal Resource Programs, 40
    Ownership and Access Issues, 42
    Cost Issues, 43
    Other Issues and Problems, 44

5   THE WASHINGTON REGIONAL PRIMATE
     RESEARCH CENTER 45
     Facilities and Programs, 46
     Ownership and Access Issues, 48
     Cost Issues, 49
     Other Issues and Problems, 50

6   THE MACROMOLECULAR CRYSTALLOGRAPHY
     RESOURCE AT THE CORNELL HIGH
     ENERGY SYNCHROTRON SOURCE 53
     User Facilities for Protein Crystallography at Synchrotrons, 53
     The MacCHESS Research Resource, 54
     Ownership and Access Issues, 56
     Cost Issues, 58
     Other Issues and Problems, 59

7   THE HUMAN GENOME CENTER:
     LAWRENCE LIVERMORE NATIONAL LABORATORY   61
     Lawrence Livermore National Laboratory, 61
     Human Genome Center, 62
     Ownership and Access Issues, 68
     Summary of Issues and Problems, 69

8   CONCLUSIONS AND RECOMMENDATIONS 71
     Features of Successful Resource Sharing, 71
     Issues and Problems, 76
     Recommendations, 84

APPENDIXES   91
A Workshop Program, 91
B Acronyms, 95

Executive Summary

The United States is entering an era of fiscal restraint, and the biomedical research community is likely to be faced with the challenge of doing more with less. One avenue that could be explored in developing the needed strategies is that of enhanced resource sharing. The public nature of science, emphasizing peer review, confirmation of results, and standardization of methods, would seem to make resource sharing a given. Independent replication provides science with quality control, and few if any laboratory experiments, or even systematic observations, can be duplicated accurately without some contact with the original author or data. Other studies may require specimens or materials obtained or created by the original author. Despite the prospect of more and more talented scientists, chasing dwindling or stagnant research funds and an increasing complexity of both clinical and basic science that would seem to demand more collaboration, a number of contemporary observers have commented on an apparent decline in the openness and willingness to share information and resources that has traditionally been viewed as a characteristic feature of science. The workshop summarized in this report was an initial attempt to examine the status of resource sharing in biomedical research, to identify existing or emerging barriers to effective sharing, and to recommend additional actions.

As an initial step in addressing the issues of whether and how to promote resource sharing, an eight-person committee with expertise in basic and clinical sciences, research administration, drug development, and public policy was charged with planning and conducting a workshop to identify some "best practices" and make the scientific public aware of the most common and most difficult problems in the area of resource sharing. The committee met in

September 1995 to plan the workshop, a task in which is was assisted by eight invited liaisons from federal agencies and scientific societies. The conclusions and recommendations of this report are however solely those of the committee.

The workshop held in Washington, D.C., on January 22–23, 1996, was built around six case studies of large-scale resource sharing, representing models of two very different institutional arrangements: "repository-type" activities and "user facilities" or centers. The resources shared by the case studies include biological materials such as whole animals, information, and instruments or equipment. By analyzing these cases in some detail, the committee hoped to better understand the roles of different institutions in influencing sharing, to identify common problems that stand in the way of effective sharing, and to suggest some approaches to their solution.

CASE STUDIES

The American Type Culture Collection

The American Type Culture Collection (ATCC) was founded in 1925 to serve as a national repository and distribution center for cultures of microorganisms. Since that time, viruses, animal and plant cell cultures, and recombinant DNA materials have been added. A private, nonprofit organization dedicated to the acquisition, preservation, authentication, and distribution of diverse biological materials, ATCC is now the largest general service culture collection in the world, preserving and providing these materials for use by qualified people engaged in science, industry, and education.

The Multinational Coordinated *Arabidopsis Thaliana* Genome Research Project

An international scientific effort that began in 1990, the goal of the Multinational Coordinated *Arabidopsis Thaliana* Genome Research Project is to understand the physiology, biochemistry, growth, and development of a flowering plant at the molecular level. The remarkable collaborative spirit of the participants has made it a successful model of scientific cooperation among about 3,000 participating scientists and scientific administrators in Asia, Australia, Europe, the Middle East, and the Americas. Two *Arabidopsis* stock centers preserve and distribute seeds, clones, and other biological materials to the large *Arabidopsis* research community worldwide. Shared databases include a comprehensive collection of many types of information; an on-line system primarily devoted to stock center operations but, like the other information systems, readily accessible to anyone with a connection to the Internet; and a database of cDNA sequences and expressed sequence tags (ESTs) that periodically sends these data to the National Center for Biotechnology Information at the National Library of Medicine. Thus, it seemed an especially appropriate case with which to examine the ingredients that facilitate the sharing of research resources.

The Jackson Laboratory

A private, nonprofit, research institution on Mount Desert Island in Maine, The Jackson Laboratory (TJL) has been the major repository for genetically characterized mice for more than 60 years. Although it is a research laboratory focused on providing new information to the scientific community through genetic research with mice, TJL is also recognized internationally as the preeminent source of laboratory mice for genetic research. TJL has developed many mutant strains of mice, which it supplies to researchers, and recently has begun accepting mice with spontaneous or induced mutations from scientists at other laboratories who wish to make them generally available. More than 1.6 million mice are distributed each year.

The Washington Regional Primate Research Center

The Washington Regional Primate Research Center (WRPRC) is one of seven regional primate research centers established by congressional mandate in the 1960s to develop nonhuman primate models and employ them to examine the underlying mechanisms and processes of human diseases. WRPRC in Seattle is part of the University of Washington Health Sciences Center. The WRPRC core staff is composed of eight doctoral-level researchers. They are joined by approximately 300 other scientists and graduate, medical, dental, and veterinary students in a wide array of research projects. The WRPRC principal breeding colony (about 350 births annually) is the Primate Field Station at Medical Lake, Washington. Additional, smaller, colonies are maintained in Russia and Indonesia. The Primate Information Center of the WRPRC maintains a comprehensive database of publications on nonhuman primates. A second database, the Primate Supply Information Clearinghouse, facilitates efficient use of nonhuman primates by collecting offers from laboratories with available primates and requests from laboratories seeking specific primates. The Tissue Distribution Program is a spin-off of the clearinghouse. It provides fresh, fixed, or frozen specimens prepared in a variety of ways to laboratories throughout the world. About 3,000 tissues harvested from roughly 200 animals will be distributed this year.

The Macromolecular Crystallography Resource at the Cornell High-Energy Synchrotron Source

The Macromolecular Crystallography Resource at the Cornell High-Energy Synchrotron Source (MacCHESS), a "user facility," provides support for the collection and analysis of x-ray diffraction data from crystals of biological macromolecules using synchrotron radiation. As such, it is a rich source of information about what works in shared facilities and where problem areas or bottlenecks exist. The overall goal of the MacCHESS research resource is to provide specialized equipment for macromolecular crystallography as well as trained support staff to assist outside users. The MacCHESS staff of two scientists, three technicians, a computer programmer, a machinist, and a secretary has established an active research program designed to advance the frontiers of synchrotron radiation research and structural biology. Collaborators obtain early access to new instruments, techniques, and methods and provide additional impetus for their development and refinement. Mature methods are made available to outside scientists who use the facility on a competitive basis. During 1995, more than 200 scientists from 45 laboratories used the CHESS facilities for macromolecular crystallography experiments.

The Human Genome Center at Lawrence Livermore National Laboratory

The last of the six case studies examined is neither a repository of scientific community property, like the ATCC, nor a center for visiting scientists, like the Cornell synchrotron. The Human Genome Center at Lawrence Livermore National Laboratory (LLNL) technically is not a shared resource at all, but a federally owned, contractor-operated research and development laboratory that has become, by default, a supplier of valuable materials and information to the international scientific community without specific funding to do so. The center staff has three tasks: creating biological resources useful for genomic research, developing instrumentation and informatics for genome research, and locating genes. The experiences of the LLNL center's scientists, as they have attempted to share locally developed instrumentation and technology, information, and biological materials with the wider scientific community, provided valuable information to the committee.

CONCLUSIONS AND RECOMMENDATIONS

Features of Successful Resource Sharing

Strong Scientific Leadership in Agencies and the Research Community

Essential ingredients in successful resource sharing are the leadership of program managers in government agencies who identify opportunities and support them; the leadership of senior scientists who establish the norm for the scientific community by example and commitment to sharing resources; the leadership of scientists who direct existing shared resources to provide quality services at moderate costs; and the commitment of scientific institutions such as universities and professional societies that develop policies to facilitate and enforce resource sharing.

Adequate Core Funding

Many repositories and centers depend on a patchwork of funding from a number of different government funding agencies, industry, and private foundations, to support research or further development of the resource, as well as user fees. Sometimes the different streams of dollars may not be available to support the core administration and quality control necessary for resource sharing.

Marketing and Advertising

Advertising, marketing, and general knowledge about the availability of a resource are essential to widespread access; many resources are not shared simply because their existence is not known to scientists who require them.

Clear Guidelines About Ownership and Access

The cases reviewed at the workshop demonstrated the value of clear guidelines concerning access and ownership, although these differ depending on the resource. Planning should include guidelines for sharing—under what circumstances and with whom data and materials will be shared. This is an essential ingredient in preventing later misunderstandings and problems.

User Fees

One important source of funding for resource sharing can be user fees. These charges help to subsidize the core operations and maintenance if those research resources that are not currently commercially viable. They also help defray the costs of functions such as authentication and quality control, which are essential, if invisible, elements of first-class science.

Clear Policies for Retaining and Discarding Data and Material

Policies for the disposition of materials and information that are no longer of value will be increasingly important as the body of resources that need to be shared continues to increase more rapidly than the funding available to support them.

Quality Control

A critical attribute of a shared resource is that the distributed resource be what it is purported to be. Similarly, mechanisms to ensure the highest-quality research at limited-access resources such as a synchrotron are essential to their ongoing success.

Well-Defined Policies for Function of Research and Service at the Facility

The balance between service and research by staff is a fundamental question to be considered by all centralized facilities designed to be resource centers for the scientific community. A shared resource is greatly enhanced by the presence of an excellent scientific staff that is conducting research to improve the resource and can ensure the quality of the materials.

Sophisticated Information Retrieval and Transfer Systems

Rapid exchange of information and widespread access to data are greatly facilitated by sophisticated information retrieval and transfer systems. Rapidly evolving information systems are transforming the way research is conducted and disseminated.

Issues and Problems

The case studies, although providing many good examples of "best practices," also provided the committee with a wealth of unresolved issues and emerging problems that any future sharing effort will have to address.

One Uniform Policy on Resource Sharing Is Not Possible

The problems of resource sharing are diverse. Therefore, the solutions will be similarly diverse. There are differences in the resources to be shared, the needs of stakeholders, and the distribution of resources that stakeholders command. In gathering the material for this report, the committee has dealt with the sharing of data, materials (including experimental subjects), and equipment. It is clear that the optimal procedures for sharing these three classes will differ in most cases. The overall guiding principle in such decisions should be scientific merit and the acquisition of information of interest to the scientific community at large.

*Incentives and Rewards for Resource Sharing
Are Not Fully Developed*

The current systems for rewarding academicians or employees in industry do not encourage sharing but rather focus on individual achievements.

- **Sharing Requires Incentives, Not Disincentives.** For academic scientists, incentives are citations or other credit for use of samples made available; another incentive is having the costs of making these samples available covered by the recipient, a third party, or one's grant. Provisions for sharing data, materials, and equipment should be built into research proposals, and the sharing activities should be included as part of the progress report when grants are being considered for renewal.

- **The Importance of Data and Material Changes Over Time.** A key clone at the early stages of an investigation may be worth trading only in an actual scientific collaboration. Later, the clone may be freely available in a public repository or distributed upon request. Finally, the clone may become archaic: it should not be kept or distributed; public repositories should deaccession it.

- **Technologies and Needs Are Evolving Very Rapidly.** Any system of incentives put into place must have sufficient flexibility to evolve as well.

- **New Definitions of "Publication" May Have to Evolve to Keep Pace With the New Electronic Information Systems.** Ways of providing credit to institutions for resource sharing must be found, or support for the scientific mission of these core activities—which benefit many—will be endangered.

*Methods For Enforcing Existing Policies on
Resource Sharing Are Inadequate*

Although some policies already exist regarding sharing, the enforcement of these policies is inadequate. Although funding agencies may have to take the lead, enforcement of these policies is most likely to be effective if done at the local (university or institution) level. The issues yet to be resolved are the actual mechanism of enforcement and how the costs involved should be paid.

*There Are Many Private and Public Stakeholders in Any Major
Resource Sharing Attempt, Often With Conflicting Goals*

Economies of scale dictate that some activities are better provided as private-sector services as long as actual costs to the users do not inhibit research. However, the issues of credit and ownership go beyond the additional constraints imposed by sharing and are badly in need of clarification and resolution. One example is the status of the research exemption from licensing for university-based investigations in a climate where universities are required by law to protect intellectual property that is potentially valuable commercially.

Who Pays and What Do They Pay for?

The issues of quality control and quality assurance for shared samples or sample repositories are of major concern, because these activities are a major contributor to the costs of institutions such as TJL and ATCC. Commercial competitors willing to employ less stringent measures on a smaller selection of resources can and do offer apparently similar products at cut-rate prices. High-quality research depends on high-quality materials, and the scientific community will have to recognize that it must pay for quality control, through subsidy if not through user fees.

EXECUTIVE SUMMARY

Regulatory Requirements and Documentation Can Be Unnecessarily Complex and Burdensome

Regulations promulgated by government agencies affect resource sharing disproportionately. The regulatory burden on ATCC for shipping biological samples and the various municipal, state, and federal regulations governing animal care and shipping are two examples.

Education of Scientists Covers Neither the Ethos of Sharing Nor Intellectual and Tangible Property Issues

During training, there is no formal emphasis on the merits of sharing or the benefits of collaborations, and in an increasingly competitive atmosphere where resources are limited, the benefits of sharing may be unappreciated.

Resource Sharing Can Have National and International Implications

Wherever resources are saturable or irreplaceable, all efforts should concentrate on viewing the scientific utility of such resources from a worldwide perspective. Procedures should be developed for worldwide review of competing applications for limited resources or facilities.

There Is a Gap in Leadership

Sharing of research resources lacks high profile leadership (for example, the president of a major scientific society or the president of the National Academy of Sciences). Academic institutions, government agencies, and industry have failed to focus the scientific community.

Partnerships in Resource Sharing May Be Unequal

The issue of fairness in access and opportunities to utilize resource sharing is ongoing, because there are typically inequities among those seeking access to saturable resources or costly resources.

Monopolies Can Be Good or Bad

Federal funding policies typically require competition for funds, but in some cases this may be an artifice that is unwarranted. The goal should be to identify the most cost effective methods and highest-quality resources for the scientific community.

RECOMMENDATIONS

Administrators of research institutions, grant administrators, scientists, and industry representatives should meet to develop policies to foster sharing of resources. These policies should explicitly address the following:

- Sources of reliable funding for provision of materials and services to the research community.
- Training and education regarding the ethos and the value of sharing and related intellectual property issues, including the merit of patents and licensing
- Rewards and incentives for researchers who share resources
- Mechanisms for enforcing agreed-upon resource sharing policies within and across institutions
- Role of the technology transfer office in facilitating resource sharing
- Current National Institutes of Health guidelines governing university-industry relationships

Federal and private funding agencies and industry should jointly undertake a suitable cost-benefit analysis and explore mechanisms to enhance the efficiency both of funding shared resources and of sharing resources.

Because of the growth of economic nationalism and to avoid unnecessary duplication, the world scientific academies should convene to identify barriers to sharing resources across national boundaries and should develop mechanisms to overcome them.

Because the private sector will continue to have a major impact on resource sharing, representatives from industry, nonprofit institutions, and funding agencies should be brought together to work toward solutions of current problems such as the following:

- Overreaching claims on future ownership of inventions by providers of shared resources and research tools
- Competition between private-sector activities and public shared resources
- How to protect the research exemption for licensed intellectual and tangible properties
- Impediments to biomedical research and education caused by confidentiality requirements

A cost-benefit analysis should be conducted to evaluate the possible impediments to resource sharing caused by government regulations.

1

Introduction

The United States is entering an era of fiscal restraint, and perhaps even more than those in clinical practice, the biomedical research community is likely to be faced with the challenge of doing more with less. This challenge may require the development of innovative strategies to facilitate and enhance the efforts of our talented scientists. One possible avenue that could be explored in developing the needed strategies is that of enhanced resource sharing. The public nature of science, emphasizing peer review, confirmation of results, and standardization of methods, would seem to make resource sharing a given. Independent replication provides science with quality control, and few if any laboratory experiments, or even systematic observations, can be duplicated accurately without some contact with the original author or data. Sometimes a telephone call will suffice; other times it may require an extended visit and hands-on training in a new technique or instrument; still other times may require acquiring specimens or materials obtained or created by the original author. Despite the prospect of more and more talented scientists, chasing dwindling or stagnant research funds and an increasing complexity of both clinical and basic science that would seem to demand more collaboration, a number of contemporary observers have commented on an apparent decline in the openness and willingness to share information and resources that has traditionally been viewed as a characteristic feature of science. The workshop summarized in this report was an initial attempt to examine the status of resource sharing in biomedical research, to identify existing or emerging barriers to effective sharing, and to recommend additional actions.

COMPETITION FOR FUNDS

The National Science Foundation (NSF) reports that total federal funding for biomedical research, in inflation-adjusted dollars, has leveled off (National Science Foundation, 1995). The current emphasis on controlling federal spending makes it unlikely that this trend will be reversed in the near future. The federal investment in nondefense research and development (R&D) is projected by the American Association for the Advancement of Science to decrease by approximately 33 percent in real terms by 2002 (Lane, 1996). Universities continue to turn out new Ph.D.s in the life sciences, however, increasing the supply about 5 percent annually (and taking on about 5.5 percent more postdoctoral appointees each year). This has resulted in declining success rates for those seeking traditional investigator-initiated research grants. Such intense competition has not affected publication, but many researchers report that the intense competition has made them think twice about sharing prepublication data, tips on laboratory technique, and important reagents (purified proteins, cloned genes, mouse strains, etc.) with potential rivals (Marshall, 1990; Cohen, 1995). Anecdotes about nonresponsiveness, incomplete sharing, and even deliberate misdirection abound (Werb, in Marshall, 1990; Rensberger, 1994). Young scientists may be tempted to hoard information and materials since unlike more senior researchers, they are most often not able to demand coauthorship or continued collaboration as a quid pro quo and are thus vulnerable to being "scooped" by a more established competitor who has more personnel and funding to exploit a new resource. Senior scientists may be reluctant to permit doctoral and postdoctoral students to take materials or even data with them to an independent position.

INCENTIVES FOR SCIENTISTS

Stiff competition for dwindling funds would seem to be an incentive for sharing, but the path to success as a scientist lies with individual accomplishments. Grants, publications, and citations are the steps on the ladder of scientific success, and even the order of the authors on a multiple-author paper can be contentious. Promotion and tenure committees often make judgments based on the number of publications authored by a particular investigator, without regard for the role played by this investigator in the overall process of science. There are few mechanisms in place actively encouraging resource sharing or reinforcing it when it appears.

Less obvious but just as important is the lack of scientific career incentives for caretakers of the common property generated by sharing. Nothing illustrates this point better than the case of Maynard Olson and his colleagues at the University of Washington, who spent two years collecting some 60,000 yeast artificial chromosome (YAC) clones. A strong belief that anything made with the support of the National Institutes of Health (NIH) belongs in the public domain initially led members of the group to encourage other investigators to send them interesting clones, for which they would try to find a match. They reported that at one point they were conducting screening for 85 different investigators and found they had no time for projects of their own. Their solution to this dilemma was to share the entire library of clones, and the associated burden of screening new clones for other investigators, with six other labs. This action could hardly be characterized as selfish, but it reveals the powerful contingencies steering even the best-intentioned scientists away from serving the larger community and toward projects of their own.

A recent trend that many feel makes this dearth of incentives for sharing especially important is the successful commercialization of much basic research in biology. The Bayh-Dole Act of 1990 and the U.S. Technology Transfer Act of 1986 contained provisions to stimulate commercial development of basic research conducted by federal agencies and their grantees by encouraging patenting and licensing agreements with private industry, which often showed little interest in developing ideas in the public domain. The incentive for the agencies and grantees is monetary—the individual scientists and their institutions are allowed to share in royalties resulting from their work. The institutions can even accept advance funding from industry partners in return for preferential access to future research findings. As Table 1–1 illustrates, the financial impact on grantees (universities) has been substantial.

TABLE 1–1 Fiscal Year 1994 Royalties Received by the Top 10 United States Universities

University	Royalties ($)
University of California (system)	50,210,000
Stanford University	37,700,000
Columbia University	26,746,141
Michigan State University	14,556,761
University of Washington	12,300,000
Iowa State University	9,600,000
University of Wisconsin	8,348,713
Florida State University	6,771,968
Harvard University	5,817,671
University of Florida	5,177,050

SOURCE: Hoffman (1995).

Blumenthal et al. (1996) provide data from the other side of the ledger. They surveyed private firms conducting or sponsoring research in the life sciences in the United States. More than 90 percent have some relationship with academia. More than half support university research. Extrapolating from their sample, Blumenthal and his colleagues estimate that private-sector companies supported more than 6,000 academic research projects in 1994, at a cost of $1.5 billion. More than 60 percent of companies investing in academic research have reported realizing patents, products, and sales as a result.

The Bayh-Dole and U.S. Technology Transfer Acts thus appear to have resulted in mutually profitable partnerships between industry and universities. Another, less propitious consequence has also been quantified by Blumenthal et al. (1996): a survey of life science companies showed that 82 percent of companies supporting research relationships with academic institutions sometimes require keeping information confidential until a patent application is filed. Nearly half of these companies indicated that their agreements with universities required academic researchers to protect confidential proprietary information resulting from company-sponsored research longer than is necessary to file a patent application. Rosenberg (1996) provides several examples, from personal experience, of secrecy in medical research, arguing that it is rapidly becoming a common and accepted practice, to the detriment of science and medicine.

NATIONALISM

A second recent trend potentially undermining the culture of sharing is a sort of "scientific nationalism" as countries seek to protect or exploit unique resources. Roughly half of all drugs in clinical use stem from a product of nature, and prospectors seeking further potential products in biota all over the world may number in the hundreds. The United Nations Biodiversity Convention of 1992 tried to ensure that profits from such products returned to the place of origin. Despite some successes, huge payoffs remain elusive. Drug companies estimate that on average, 10,000 to 100,000 substances are screened for every profitable drug brought to market. One common result in developing nations however has been resentment and anger toward bioprospectors from industrialized countries, who are suspected of circumventing the convention. In response, several countries have passed laws severely restricting export of native flora and fauna, regardless of the intended use.

METHODS AND GOALS OF THIS STUDY

A Member Survey

Several previous National Research Council (NRC) reports have touched on some of the issues noted above, for example, *Sharing Research Data* (Feinberg, et al., 1985) and *Sharing Laboratory Resources: Genetically Altered Mice* (National Research Council, 1994), so the current project began with an informal survey of members of the Institute of Medicine (IOM) and of relevant sections of the National Academy of Sciences (NAS). The survey inquired about members' own difficulties in resource sharing, what kinds of resources could and should be shared, what the scope and mechanisms of such sharing might be, and what specific examples of successful or failed efforts at resource sharing would be worth examining in detail.

Responses from NAS and IOM members made it clear that any study of resource sharing would have to recognize the multiple meanings of both "resource" and "sharing." The former, for example, might encompass biological **materials** (tissue samples, cell lines, bacteria, viruses, antibodies, genes or gene fragments, and plasmids); **information** (data, databases, patient registries, or recipes and procedures); **instrumentation** (microscopes of various sorts, synchrotrons, accelerator or magnetic resonance spectroscopes, and other expensive equipment); and **experimental subjects** (primates, mutant strains of mice or fish, patient registries, or families with known or suspected genetic diseases). Each type of resource presents special considerations for sharing, though all will have to address the costs of production and distribution or the responsibility for maintaining the shared resource. Biomedical, behavioral, and epidemiological data vary in their content, level, form, and structure. Distinctions between the materials and the data themselves are often blurred (Sieber, 1990).

In addition, "sharing" could involve large- or small-scale **collaborations** within or across institutions; **scientist-to-scientist exchanges**; deposition of resources into regional or national **"public domain" repositories**; or **"time-sharing" of rare or expensive facilities** among collocated staff and visiting "users." Adding still further complexity is the network of interrelationships among the many actors influencing scientists' decisions about when, where, what, how much, and with whom to share (see Figure 1-1).

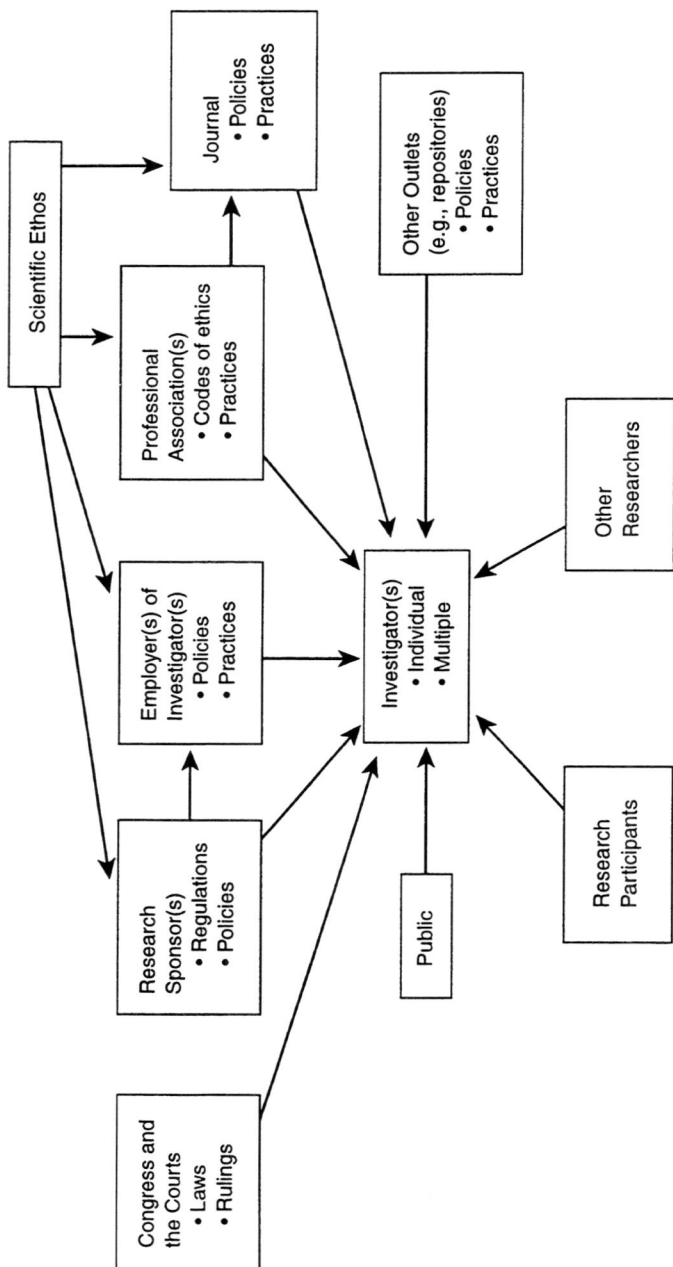

FIGURE 1-1 Potential influences on investigator's decisions to share. SOURCE: Cordray et al. (1990).

The Committee

An eight-person committee with expertise in basic and clinical sciences, research administration, drug development, and public policy was charged with planning and conducting a workshop to identify some "best practices" and make the scientific public aware of the most common and most difficult problems in the area of resource sharing. Specifically, the workshop was to (1) review the current status of sharing in a few particular categories of biomedical resources; (2) identify existing programs, initiatives, and mechanisms in place for sharing these resources; (3) identify future needs, obstacles, and strategies that will promote sharing; and (4) assess agreement within the biomedical research community and relevant funding agencies about the need for advice and recommendations in these areas. The committee was joined by eight representatives from federal agencies and scientific societies in a September 1995 meeting to plan the workshop.

The Workshop

The workshop, held in Washington, D.C., on January 22–23, 1996, was built around six case studies of large-scale resource sharing, representing models of two very different institutional arrangements: "repository-type" activities and "user facilities" or centers. (See Appendix A for the program.) The resources shared by the case studies include biological materials such as whole animals, information, and instruments or equipment. By analyzing these cases in some detail, the committee hoped to identify common problems that stand in the way of effective resource sharing, to better understand the roles of different institutions in influencing sharing, to highlight the advantage of sharing for the scientific community, and to stimulate support for sharing from that community.

Each presenter was asked to describe the relevant activity or facility, and to specifically address the operations of the activity or facility in terms of the following:

- How is the issue of *ownership* addressed? Do the contributors maintain any control over the materials, their distribution, or use? For how long? Do they get credit of any sort, either with the facility or with the scientific community? If not, what is their incentive for contributing? Are any conditions imposed on contributors, (e.g., provide documentation of agreement among all members of a collaboration)?
- Who can *access* the shared materials, and how? What mechanisms are employed for disseminating information on availability? Are there any

conditions or restrictions on use? Any rules for acknowledgment of the original contributor?

- What is the primary *function* of the facility? R&D? Distributor of R&D tools and products? Curator? What is your criterion for success? Is the endgame a steady state, or do you foresee a time when the facility, or some functions of the facility, will no longer be necessary? If so, how will you know when that time has arrived? Do you have any plans for disposition of resources or functions in the event the facility has to cease operations involuntarily?
- What are the *costs* (nonmonetary as well as monetary) associated with maintaining the shared resources, and how are they covered? What kind of quality control process is employed? Are there financial incentives for contributing or using shared materials? Barriers?
- What *other* issues or problems create difficulties for your facility? How would you prioritize among all of these issues?

The Report and Its Recommendations

This report is a distillation of the resulting talks on the case studies; additional presentations on the roles of government, professional societies and journals, and private industry; and discussions of invited guests from the public, nonprofit, and private sectors. The committee is however solely responsible for the conclusions and recommendations of this report.

REFERENCES

Blumenthal, D., Causino, N., Campbell, E., and Louis, K.S. 1996. Relationships between academic institutions and industry in the life sciences—An industry survey. *New England Journal of Medicine* 334(6):368–373.

Cohen, J. 1995. The culture of credit. *Science* 268(June 23):1706–1711.

Cordray, D.S., Pion G.M., and Baruch, R.F. 1990. Sharing research data: With whom, when, and how much? Paper presented at the Public Health Service Workshop on Data Management, Access, Sharing, and Retention in Biomedical, Behavioral, and Epidemiological Research, April 25–26, 1990, Chevy Chase, Maryland.

Feinberg, S.E., Martin, M.E., and Straf, M.L., eds. 1985. *Sharing Research Data*. Washington, D.C.: National Academy Press.

Hoffman, D.C. 1995. The AUTM Licensing Survey. Norwalk, Conn.: Association of University Technology Managers.

Lane, N. 1996. Thin ice over deep water: Science and technology in a 7-year downsizing. Presentation at the American Astronomical Society Meeting, January 15, 1996, San Antonio, Texas.

Marshall, E. 1990. Data sharing: A declining ethic? *Science* 248:952–957.

National Research Council. 1994. *Sharing Laboratory Resources: Genetically Altered Mice.* Washington, D.C.: National Academy Press.

National Science Foundation (NSF). 1995. *National Patterns of R & D Resources: 1994.* Arlington, Va.: NSF/Division of Science Resource Studies, p. 8.

Rensberger, B. 1994. Era of transition: Successful science, troubled scientists. *Journal of NIH Research.* (August 6):29–31.

Rosenberg, S.A. 1996. Secrecy in medical research. *New England Journal of Medicine* 334(6):392–394.

Sieber, J.E. 1990. Investigator's concerns about data sharing. Paper presented at the Public Health Service Workshop on Data Management, Access, Sharing, and Retention in Biomedical, Behavioral, and Epidemiological Research, April 25–26, 1990, Chevy Chase, Maryland.

2

The American Type Culture Collection

The American Type Culture Collection (ATCC) is a private, nonprofit organization dedicated to the acquisition, preservation, authentication, and distribution—the "APAD" activities—of diverse biological materials. ATCC was founded by scientists in 1925 to serve as a national repository and distribution center for cultures of microorganisms. Since that time, viruses, animal and plant cell cultures, and recombinant DNA materials have been added. ATCC is now the largest general service culture collection in the world, with collections in six areas: Bacteriology, Cell Culture, Molecular Biology, Mycology, Protistology, and Virology.

The mission of ATCC is to serve as the world's leading repository for standard reference cultures, related biological materials, and associated data. ATCC provides for the permanent preservation and availability of these materials for use by qualified people engaged in science, industry, and education. In pursuit of its mission, ATCC's principal goals are

- to acquire, preserve, propagate, and distribute cell cultures, microorganisms, viruses, cellular products, and biological materials used in and derived from recombinant DNA technology;
- to maintain the highest standards of authentication, documentation, and maintenance of the characteristics and viability of the materials entrusted to the collections;
- to pursue research based on or related to the collections;
- to provide the highest-quality service to members of the scientific, commercial, and public sectors who work with collection materials;

- to educate scientists and the public about ATCC holdings and activities via training programs, lectures, publications, databases and other means; and
- to collect, manage, disseminate, and exchange information applicable to the materials in the collections.

ATCC is affiliated with 22 professional scientific organizations, the primary users ofits cultures and services. ATCC policies are determined by a 15–member board of directors composed of representatives from these organizations and the community at-large.

GENERAL FACILITIES

ATCC employs a staff of 220 individuals. The facility is presently located in Rockville, Maryland, on approximately 5 acres of land. The Carlson building (53,000 square feet) was designed and equipped specifically for the study and maintenance of cultures. It houses the six collections, a library, conference/seminar areas, a workshop laboratory, a greenhouse, and Manufacturing. Two other buildings house Sales and Marketing, Shipping, Information Services, and the administrative offices.

ATCC animal facilities are accredited by the American Association for Accreditation of Laboratory Animal Care and registered by the U.S. Department of Agriculture (USDA). The greenhouse and all laboratories in which plant pathogens are handled are inspected by state and federal (USDA Animal and Plant Health Inspection Service/Plant Protection and Quarantine) officials for compliance with quarantine regulations. Buildings have restricted access and are monitored 24 hours a day. In emergency situations, an auxiliary generator supplies power to freezers, refrigerators, and other critical instruments. For added security, a duplicate supply of all freeze-dried material is stored in Blacksburg, Virginia; backup liquid nitrogen storage for frozen material is located in Frederick, Maryland.

The ATCC scientific programs are supported by Manufacturing, Sales and Marketing, Bioinformatics, and Publications. Manufacturing occupies almost 7,000 square feet of space. Its staff assists in the freezing and freeze-drying of cultures and maintains the culture inventory. Presently there are more than 500,000 ampules of bacteria and fungi and 400,000 ampules of virus antisera stored in walk-in cold rooms at $+4°C$ and $-20°C$ and 68,000 ampules of viruses stored in mechanical freezers at $-70°C$. More than 500,000 vials of cell lines, protists, and seed material for other collections are stored in vacuum-insulated freezers cooled with liquid nitrogen at $-196°C$.

Sales and Marketing uses a computerized inventory, order processing, and invoicing system to provide customers with current information on usage, availability, and replacement of cultures. Bioinformatics assists in developing computerized data management systems. Publications is responsible for production of catalogues, newsletters, technical manuals, and other informational brochures.

PROGRAMS

ATCC has four program areas focused on the development and distribution of bioscience products and services, bioscience research, and technology transfer.

Collection, Research, and Services Program

The Collection, Research, and Services (CRS) Program is responsible for ATCC's primary mission of maintaining and providing the world's largest and most diverse collection of biological cultures and culture-derived materials. For 70 years, scientists throughout the world have donated biological materials to ATCC. Those to be accessioned are selected by the collection managers and CRS program directors. Although APAD activities differ slightly according to the type of material, the general criteria for accession are the same and include historical significance, amenability to preservation, level of characterization, and value to the scientific community.

Before being accessioned and catalogued for distribution, biological material is subjected to a series of tests to check viability, purity, identity, preferred temperature and medium for growth and/or sporulation, and methods of preservation. For this reason, some have referred to ATCC as a de facto bureau of standards in biology (a field without an official bureau of standards). No fee is charged for deposits accepted into the collection, and no cultures are purchased from investigators. A depositor has a lifetime right to secure a culture of that deposit without charge.

All collections use the seed stock system to maintain their distribution stock. As each deposit is accessioned, some ampules are set apart as seed stock and others are designated as order stock. When the order stock becomes depleted, an ampule of seed stock is opened and new specimens are prepared from it and freeze-dried or frozen as new order stock. The seed stock is always the closest material available to the original deposit.

ATCC currently has more than 80,000 items catalogued and available for use by the scientific community. Collection materials are growing exponentially with the addition of cDNA clones from the Institute for Genomic

Research and the Integrated Molecular Analysis of Genome Expression Consortium (IMAGE). The material currently available consists of the following:

Bacteriology (bacteria and bacteriophages)	15,203 items
Cell Culture (cell lines and hybridomas)	3,403 items
Molecular Biology (recombinant DNA materials)	31,339 items
Mycology (filamentous fungi and yeasts)	27,370 items
Plant tissue cultures	76 items
Seeds	97 items
Protistology (protozoa and algae)	1,330 items
Virology (plant viruses and antisera)	1,010 items
Animal viruses, chlamydiae, rickettsiae, and antisera	2,485 items
Total	81,303 items

ATCC distributes cultures for a fee to scientists and educators worldwide who have the appropriate documentation. Prices reflect the ATCC cost of preparing, testing, preserving, maintaining, and shipping cultures or reagents. ATCC complies with all domestic and international regulations and guidelines for packaging, labeling, and transporting infectious substances and potentially infectious materials. The packaging and labeling requirements of the U.S. Postal Service, Department of Transportation, and Public Health Service for domestic shipments, and International Air Transport Association requirements for international shipments, are followed. ATCC also works closely with other agencies, such as the USDA and the Department of Commerce, as well as the Public Health Service, to obtain all required permits and export licenses.

In the last 15 years, a total of 1,133,945 items have been distributed. The following listing of annual totals reveals a recent slowing and even a reversal of the steady growth that characterized most of this period:

1980	—	36,846	**1988**	— 105,531
1981	—	40,740	**1989**	— 118,413
1982	—	47,642	**1990**	— 127,398
1983	—	60,144	**1991**	— 134,043
1984	—	67,714	**1992**	— 152,809
1985	—	71,631	**1993**	— 151,475
1986	—	78,794	**1994**	— 139,245
1987	—	92,240		

Distribution figures for 1994 indicate that about one-third of ATCC materials are distributed to foreign countries for use in clinical, industrial, research, university, and government laboratories, with industrial labs (50 percent) and university labs (25 percent) accounting for the bulk of demand both here and abroad.

CRS program directors and collection managers routinely examine collection holdings to determine their relevance to the scientific community. Taxonomically significant strains must be retained. All others are reviewed periodically for possible "deaccessioning" or discarding. The same criteria used for accessioning biological material are used for deaccession.

Each collection has an advisory committee composed of external scientists with recognized expertise in many disciplines that meets regularly with ATCC staff to provide advice and assistance in acquisition and authentication of materials.

Professional Services Program

The Professional Services Program provides several products and services that complement collection activities and expertise at ATCC. Established in 1949 as a depository for strains that were cited in U.S. patents, ATCC was designated in 1981 as the first International Depository Authority under the International Budapest Treaty for biotechnology patents. In addition to its patent deposit service, ATCC offers "safety deposit," a proprietary storage for customers.

Contract laboratory services include a variety of standard and custom services in the areas of cell culture, molecular biology, microbiology, and others. Products are offered for propagating, testing, and preserving cultures.

Education Services Program

The Education Services Program provides training programs for the biological sciences. Conferences and courses are arranged by ATCC in direct response to needs identified by the collection staff or an outside source. Subjects include quality control measures; managing strain data; obtaining patents in biotechnology; and identifying, preserving, and maintaining cultures. The workshop program provides hands-on laboratory experience in areas such as cytogenic technology, diagnostics, fermentation microbiology, recombinant DNA technology, hybridomas and monoclonal antibody technology, hybridoma data management, and DNA sequencing and polymerase chain reaction (PCR) technology. Teaching kits and videos have been prepared by several ATCC scientists.

Many of the ATCC staff with expertise in specialized areas are available for consulting work. They can recommend strains for specific uses; preserving, packaging, and shipping techniques; laboratory practices and quality control procedures; and recording, managing, and administering nonclinical experiments. ATCC scientists can often provide specialized bibliographies to outside investigators. Sponsored visiting scientists are welcome to conduct research of mutual interest.

Information Services Program

The Information Services Program maintains the extensive databases of biological information developed and stored at ATCC. Information on material is gathered and updated through direct contact with depositors and computer-based literature searches. Data from accession forms and reprints are stored by means of database software. Reprints are transferred to microfiche files. Computer database information, backed up and stored off-site as a safety precaution, is retrieved for reports, product sheets, and catalogues.

Each collection issues a catalogue of its holdings in hard copy every three to four years. Catalogues are concise compilations of the data and literature references of greatest interest to users. They are also available in electronic form in CD-ROM and PC diskette versions and on-line via the Internet. The catalogues are widely publicized by news releases and announcements in ATCC newsletters. Previously distributed free of charge, there is now a charge for new editions to cover printing and mailing costs.

The *ATCC Quarterly Newsletter,* distributed free to about 15,000 scientists, lists all new materials and publicizes other important collection and organizational news. ATCC also publishes technical manuals on quality control measures, freezing and freeze-drying, packaging and shipping of biological materials, and specific uses of ATCC strains. The catalogues and manuals are regarded as general reference documents.

The Bioinformatics section provides information to the scientific community through on-line systems and is involved in establishing international networks of microbial and cell line information resources. On-line access to collection databases is available via Internet Gopher server, the World Wide Web, the Microbial Strain Data Network (MSDN), and the World Data Center (WDC). At the present time, Bioinformatics is working to improve the availability and usefulness of collection information through an integrated scientific database (ISDB) with a centralized bibliographic reference system and a standardized terminology and synonym resource. The integrated system not only will facilitate the identification of strains with specific characteristics, regardless of which collection holds the strain, but also will enable ATCC to efficiently update and modify information established after the strain was first deposited.

OWNERSHIP AND ACCESS ISSUES

The question of who owns the materials in the ATCC collections has recently been the impetus for a detailed explication of ATCC acquisition policy. Over the course of 70 years, donors have provided materials in a variety of ways. Some simply made gifts. Others gave only with very explicit restrictions. Individuals sometimes gave without approval by their institution, and institutions sometimes gave without sign-offs by the investigators. Sometimes one investigator gave without checking with his or her coinvestigators. As a result, most materials in ATCC, even if cited in valid patents, expired or invalid patents, abandoned patent applications, and pending patents—if not restricted by the applicant—have open access and open use. The rest of the material is in what are called special collections; they contain restricted access materials, unreleased patent cultures, safety deposit material, intramural R&D materials, materials from extramural partnerships, and from technology transfer materials. Thus, the categories of resources at ATCC in terms of intellectual property are free access (public), limited access (public), or limited access and use in the restricted category.

In January 1996, in response to the increasingly frequent desire of potential donors to hold on to their materials or seriously limit their distribution, until the commercial value becomes clear, and in order to limit ATCC liability in intellectual property disputes, ATCC issued detailed policies covering all cultures acquired after that date:

1. For single cultures or small numbers of related cultures, ATCC prefers that potential donors contribute cultures to one of the general collections in a gift format without any donor-imposed restrictions on access or use. ATCC thus acquires rights to use, propagate, and distribute the culture(s) to customers for a fee. In exchange, ATCC accepts the responsibility for authentication and preservation of the material.

2. Any purchaser of such donated cultures (including ATCC) can use them to discover and develop new patentable products or processes. It is the responsibility of the purchaser to determine whether the new products or processes infringe the intellectual property rights of other parties.

3. Should option 1 not meet the depositor's requirements, ATCC will discuss the option of deposit in a general collection with depositor-requested access or use restrictions. If a deposit is accepted with such restrictions, ATCC will communicate those restrictions to potential purchasers through catalogue records and product sheets.

4. Any purchaser of such "depositor-restricted" cultures (including ATCC) should do so in conformance with those restrictions. It is the responsibility of the purchaser to determine whether any new products or processes developed using these cultures infringe on the intellectual property rights of the depositor or other parties.

5. ATCC will offer potential depositors of large numbers of related biological materials the opportunity of establishing a special collection. This is ATCC's preferred option. ATCC will help the potential depositor identify sources of funds to support such collections, but the depositor will arrange to endow the special collection with sufficient funds for its long-term support. Special collections may contain donated and/or donor-restricted materials.

6. ATCC will continue to offer, for an annual fee, two additional forms of deposit under which access or use restrictions are permitted. These are the Patent Depository, for use in conjunction with a pending patent application, and Safe Deposit, whereby ATCC maintains the materials without advertising.

7. Should none of the options described above meet the needs of originators of desirable materials, ATCC may propose alternative arrangements such as contracts, joint ventures, partnerships, or other means of working with the originator to develop new products, processes, or services. This may include licensing arrangements, with the right of sublicensing to third parties, or agency arrangements whereby technology transfers are made on behalf of the originator.

It remains to be seen what effect these policies and options will have on donations, but ATCC's clear delineation of conditions and strong affirmation of the tradition of unrestricted sharing are certainly welcome. Not covered by these new policies, but clearly a problem for ATCC, is the issue of appropriate credentials for purchasers. At the moment, a knowledgeable-sounding request on institutional stationery appears sufficient for most purchases. ATCC, understandably, does not want to become an enforcer, but a mishap last summer in which three vials of *Yersinia pestis* (the agent of bubonic plague) were shipped to an individual in Ohio who fraudulently portrayed himself as a legitimate scientist has made it clear that more stringent criteria are necessary for at least clearly hazardous biological materials.

COST ISSUES

No collection of living germ plasm has ever become financially self-sufficient. It is common practice for the user community to partially absorb the cost of curatorial functions. The major portion of ATCC's funding is provided by culture fees. The fees charged for materials are not directly related to the total costs involved in the authentication and documentation of materials being accessioned into the collection, studies on long-term preservation methods, maintenance and long-term storage expenses, and the cost of distribution. Eighty percent of ATCC collections do not generate any revenue.

While the demand for collection services has been increasing, the conventional areas of financial support for ATCC have been steadily decreasing. Federal support for collection activities dropped to 16 percent in 1995, and federal money for infrastructure has disappeared completely. Responsibility for covering this shortfall has been transferred to ATCC and its users as summarized in the following table:

TABLE 2-1 ATCC Revenues, 1993–1995

Year	Total Revenues	% of Total Revenues		
		Culture and Service Fees	Grants and Contracts	
			APAD	Research
1993	16,127,000	71.53	24.50	3.97
1994	16,934,000	75.95	19.84	4.21
1995	17,932,235	76.44	16.57	6.99

NOTE: APAD = acquisition, preservation, authentication, and distribution.

Not apparent from the table is the fact that ATCC increased prices sharply between 1989 and 1994, and probably cannot continue to do so without a negative effect on sales. Private collections and commercial repositories are already a significant source of competition, the latter doing so by "cherry-picking." That is, they are maintaining and distributing only those materials for which there is a heavy current demand, and are ignoring the less popular materials that comprise 80 percent of ATCC's collections and impose significant additional cost on ATCC operations. ATCC leadership believes that maintaining these "unprofitable" cultures is an indispensable part of its mission, not simply because some may later prove useful (the bacterium *Thermus aquaticus* was discovered years before its extraordinary heat resistance made it the key to PCR and the explosive growth of biotechnology), but also because biology as a science depends on access to a wide variety of well-characterized specimens.

Although the committee agrees with this position in principle, it is obvious that no modern-day Noah can aspire to maintain a representative of every living organism. ATCC itself has acknowledged this in its new policies regarding acquisition of orphan collections and other special collections, and concedes that some deaccessioning policy and procedures are badly needed.

OTHER ISSUES AND PROBLEMS

Related to both the increasing competition and the need for a deaccessioning policy are some issues of international relations that must be addressed before their adverse effects on microbiology become irreparable. Prominent among these is the growth of foreign culture collections which are totally subsidized by foreign governments. In many instances, ATCC is asked to help stock these collections. The question arises as to whether foreign collections will be equally available to citizens of the sponsoring country and to scientists from other countries. The time is past when only the United States was capable of establishing and maintaining a first-class repository for biological materials, and it is time for the international scientific community to take advantage of this fact rather than squander funds in unnecessary duplication. ATCC may be the largest and most diverse collection in the world, but it is not the largest and most diverse in every area. The German national collection, for example has just announced it will be funding 77 scientists with long-term support for one of the best mycology collections in the world. ATCC currently has three Ph.D. mycologists. A precedent for the sort of international agreement required already exists in the Budapest Treaty governing patent deposits, 35 countries are signatories to this treaty, the most important point of which is the agreement to recognize deposits made in any of 28 international depository authorities (IDAs).

A different and more difficult international issue arises from the belief increasingly expressed by developing countries that indigenous germ plasm is being appropriated unfairly by the developed nations and serving as the basis of lucrative commercial enterprises. The result has been a plethora of national policies restricting export of indigenous materials and establishing highly proprietary national collections, even to the point of renaming organisms.

3

The Multinational Coordinated *Arabidopsis Thaliana* Genome Research Project

Arabidopsis thaliana is a noncommercial member of the mustard family that has become widely used as a model plant because it develops, reproduces, and responds to stress and disease in much the same way as many crop plants. The plant has a number of features that make it ideal for research purposes—it is easy and inexpensive to grow and produces many seeds, which is useful for genetic experiments. One especially attractive feature is its small genome (100 megabases), which simplifies and facilitates genetic analysis.

PROJECT ELEMENTS

The Multinational Coordinated *Arabidopsis Thaliana* Genome Research Project is an international scientific effort that began in 1990. Its stated goal is to understand the physiology, biochemistry, growth, and development of a flowering plant at the molecular level. The project developed when several program managers at the National Science Foundation (NSF), recognizing that research on *Arabidopsis* was accelerating, convened a series of international workshops of leading scientists to devise a long-range plan. The resulting project plan called for genetic and physiologic experiments to identify, isolate, sequence, and understand genes; the establishment of worldwide electronic communication among laboratories; the establishment of resource centers for collection and dissemination of genetic stocks, genes, and related materials; and the creation of databases so that new knowledge would be shared. The project plan also contained mechanisms for formal, annual progress reviews and periodic establishment of new goals by a multinational steering committee of leading *Arabidopsis* researchers. In this multi-institution project, or collection of related projects, NSF supported the early collaboration and planning efforts, but the U.S. scientific community now is also supported by the National Institutes of Health (NIH), the Department of Agriculture, and the Department of Energy. Ongoing communications among scientific administrators, the scientific community, and the national and international steering committees facilitate the identification of needs, rationalization and prioritization, and negotiations with agencies around resource requirements. The remarkable collaborative spirit of the participants has made it a successful model for scientific cooperation among several thousand participating scientists and scientific administrators in Asia, Australia, Europe, the Middle East, and the Americas. Thus, it seemed an especially appropriate case with which to examine the ingredients that facilitate the sharing of research resources.

Perhaps most central to the issue of sharing research resources are the biological resource centers and the informatics that facilitate exchange of information and materials. The *Arabidopsis* stock centers were established in 1991 to preserve and distribute biological materials supporting the large *Arabidopsis* research community. There are two such centers—the *Arabidopsis* Biological Resource Center (ABRC) at Ohio State University in Columbus, Ohio, and the Nottingham *Arabidopsis* Stock Centre (NASC) at the University of Nottingham, United Kingdom. Both of these stock centers have a comprehensive collection of seeds and clones as well as other research tools such as T-DNA lines and transposable element-transformed lines, transposon lines, promoter trap lines, recombinant inbred populations, and yeast artificial chromosome (YAC) and phage libraries—which they distribute worldwide. The number of stocks sent has increased significantly in the last three years, from 15,000 total seed stocks distributed in 1992 by ABRC and NASC combined, to about 45,000 seed stocks distributed in 1994. As for DNA, 1,000 clones and 6 YAC libraries were sent in 1991; just two years later, about 3,100 clones and 166 libraries were distributed, according to the NSF's Multinational Coordinated *Arabidopsis Thaliana* Genome Research Project Progress Report for Year Four. Centers are now providing considerable technical services such as multiplexed libraries to facilitate screening for specific genes.

Three major databases are key resources for sharing information. These include the Stanford-based *Arabidopsis thaliana* Database (AtDB) previously at Massachusetts General Hospital, where it was called An *Arabidopsis thaliana* Database (AAtDB). This is a comprehensive collection of many types of information, including genetic map information obtained directly from investigators or from publicly available collections and databases. The *Arabidopsis* Information Management System (AIMS) is an on-line database system running on a central machine at Michigan State University. It is devoted primarily to stock center operations, but like the other information systems, it is readily accessible to anyone with a connection to the Internet. The third major database, devoted to cDNA sequences and expressed sequence tags (ESTs), is maintained at the University of Minnesota, which periodically sends these data to the National Center for Biotechnology Information at the National Library of Medicine. In addition, several new databases have recently been developed for managing information on EST contigs (the Institute for Genomic Research) or for information on YAC contigs (University of Pennsylvania John Innes Center). The relative ease with which a World Wide Web (WWW) server can be established is leading to rapid proliferation of specialty databases.

OWNERSHIP AND ACCESS ISSUES

At this time, the U.S. stock center and databases do not accept deposits that place restrictions on materials, a policy that has in a few instances, impeded accepting some important collections. However, NASC has accepted a collection of insertional mutants in which users are required to sign a material transfer agreement that cedes commercial rights to the investigators that produced the collection.

Curators aggressively solicit materials. Whenever a paper is published, authors are sent a note requesting the materials in the paper (in the future, because obtaining deposits is such a time-consuming but important process, members of the research community, rather than members of the stock center, will solicit deposits). Quality control is also conducted by the curators. Peer pressure, the example of prominent scientists, and recognition for contributing stocks all help foster continued contributions to the centers and their associated databases. The national steering committees, originally ad hoc but now elected (the six-member American committee has two members replaced each year through e-mail balloting), wield considerable influence in this respect, as do the heads of the major laboratories, who have encouraged openness and sharing by clear public acknowledgments to depositors of data and materials. Although at this time the *Arabidopsis* community requires that genome sequences be deposited in the public database three months after they are publicly available, the multinational steering committee is considering requesting that journals publishing in this area require an accession number from the stock centers indicating that experimental materials have been deposited.

There is also no continuing ownership of materials in these stock centers (i.e., once there, they are owned by the stock center). The stock centers and databases are extremely successful because resources and information are so freely and willingly shared. As soon as raw sequence data are obtained from the Michigan State University cDNA sequencing project laboratory, for example, they are sent directly to the University of Minnesota, where the initial analysis takes place and the result are deposited in the public database. At the same time, the clones are deposited in the stock center and thereby made available to anyone interested. An interesting feature of the U.S. stock center database is that all requests to the stock center are logged on the database, which is available on-line to anybody in the world. This way anyone can find out the names of the people and the labs requesting a specific seed or clone and the date of the request. Although at first there was considerable concern that large laboratories might gain an edge over smaller ones through such information sharing, the mechanism has instead been found extremely useful in developing collaborations rather than stimulating competition.

Products of *Arabidopsis* sometimes stimulate commercial interest, and patenting is both common and encouraged, although there seems to be a strong feeling in the *Arabidopsis* community that nobody should patent genes in this organism (as opposed to a novel use of them). One consequence of this view has been a strong pressure to get sequence data, especially ESTs, into the public domain quickly, so that patenting based merely on sequence information becomes difficult or impossible.

In other cases—for example, novel applications—relevant materials and information are not published or deposited until after the patent application is filed, but once this is accomplished there has been a general commitment to sharing the resource. For example, the project's "rule" is that the sequences appear in a public database three months after they are available, and although undoubtedly the odd patent may be written on these sequences before the three months is up, this is the rule that the community itself wrote at a workshop sponsored by NSF. Another aspect of the enforcement question is a second rule, this one requiring sequencing groups to have a member of the national steering committee on the executive committee overseeing their sequencing operation, thus making it difficult for a sequencing lab to keep results secret for very long. There have been no real tests of the consequences for breaking either of these rules, but the assumption is that NSF will discontinue funding if there is a complaint from the community.

Enforcement is more complicated at the international level. The international steering committee is trying to negotiate the contribution of a collection of mutants made by a consortium in Europe, where strong pressure is being put on scientists by their funding agencies to limit distribution to those willing to cede or share future commercial benefits.

COST ISSUES

In the United States, the Department of Agriculture, the Department of Energy, the National Institutes of Health, and the National Science Foundation collectively supplied $7.5 million for *Arabidopsis* research in 1990 and $22 million in 1993. Of that total, the amount devoted to the Multinational Coordinated *Arabidopsis Thaliana* Genome Research Project over the last five years comes to about $4.2 million: $2.2 million for establishing and maintaining the various databases and $1.9 million for the stock center at Ohio State (ABRC).

Universities are unquestionably subsidizing the enterprise, but the extent and cost-effectiveness of this approach are not known. International components receive support from their own governments and the European Community.

The relatively modest amounts required by this project appear to the committee as money well spent, and unlike some of the other case studies examined, current and projected funding appears adequate. One reason for this seems to be that the services provided appear to be viewed by both the NSF and the *Arabidopsis* research community as legitimate objects of research support. That is, essential support for researchers as a group is seen as no less deserving of research dollars than the projects of individual scientists.

OTHER ISSUES AND PROBLEMS

In his presentation at the workshop, Chris Somerville of Stanford and the Carnegie Institution, one of the project's original organizers, identified factors that are instrumental in the sharing of resources. These include the leadership of program managers in government agencies that support the research; the leadership and example of senior scientists and prominent laboratories; an oversight committee with broad representation of countries and scientists that sets policy, adjudicates problems, and can make proposals to funding agencies based on the needs of the community; investment in infrastructure such as stock centers and information databases; support for workshops and other scientific meetings; and a process for annual updating of a plan. Peer pressure to share information and materials and aggressive solicitation of stocks for the centers are also important.

Among the problems identified by Chris Somerville is the requirement for U.S. funding agencies to set up stock centers and databases via a competitive process even when the steering committee could locate only one interested and capable bidder in the community. Other problems include providing stocks and services to an international community with limited funds from U.S. agencies;

as well as the ongoing administrative load imposed by the need for an active process of soliciting deposits, a time-consuming activity given the pace of research on *Arabidopsis*. More information on this case study is available from NSF in the project's progress report for year four.

4

The Jackson Laboratory

For more than 60 years, The Jackson Laboratory (TJL), a private, nonprofit, research institution on Mount Desert Island off the coast of Maine, has been the major repository for genetically defined mice. TJL is internationally recognized as the preeminent source of laboratory mice. The laboratory describes its mission as

1. providing new information to the scientific community through basic genetic research using mice;
2. providing the essential genetic resources for other scientists to do that research throughout the world; and
3. educating the next generation of scientists to carry out this work.

TJL is governed by a Board of Trustees that includes both scientists and nonscientists; a Board of Scientific Overseers; and a Director, who is also provided advice in different areas by staff scientists on four standing committees. Support for the laboratory's activities comes from a combination of federal agencies (National Institutes of Health [NIH] and National Science Foundation [NSF]), other health organizations (Howard Hughes Medical Institute, American Cancer Society, American Health Association, Cystic Fibrosis Society, Multiple Sclerosis Society, March of Dimes, and others), and fees from services or sales of laboratory mice. The total operating budget for TJL for 1995 was approximately $45 million, and about half of that is related to maintaining and distributing animal resources (mice) and related services, that is, production and sale of specific mouse mutants, maintenance of selected breeding stocks, derivation of selected strains or congenic production, surgical manipulations, embryo conservation, and bioinformatics.

ANIMAL RESOURCE PROGRAMS

Production, Sale, Derivation, and Maintenance of Mice

TJL develops mutant strains of mice as well as accepting mice from scientists who wish to make them available to other scientists. These mice from external sources may be spontaneous mutations, or they may be induced mutations. The laboratory also maintains pedigreed stocks of mouse strains using breeding programs that are designed to ensure their genetically unique qualities. Mice that are accepted by TJL must all be cesarean rederived to ensure that they are disease free. At this point they become Jackson Laboratory (JAX) mice and are distributed as such according to TJL policies. Internally, TJL has divided its 1,800 or so strains of mice into seven categories, each managed in a separate subunit known as a resource (instead of a department or division):

1. Induced Mutants—these include transgenics, induced and targeted mutations;
2. Mouse Mutants—spontaneous mutations;
3. Special Mouse Stocks—congenic and recombinant inbred strains;
4. Foundation Stocks—pedigreed source colonies for inbred strains;
5. Individual Research Colonies—these include all of the above types;
6. Animal Resources—the expansion and production of colonies of inbred, mutant, and special strains in high demand; and
7. Frozen Embryos—all of the above types.

The largest numbers of mice are distributed from the production colonies (Animal Resources) (1.6 million annually); the Induced Mutant, Mouse Mutant, and Special Mouse Stocks Resources each distribute 10,000–12,000 mice annually.

The newest and fastest-growing resource is the Induced Mutant Resource (IMR), which may include about 235 strains at any one time. Until this resource was initiated in 1992, TJL distributed only mice developed by its in-house research staff. Almost all mice in the IMR originate outside TJL. The original plan for this resource was that approximately 50 percent of the strains included would be requested from authors of published papers and about 50 percent would be offered by external scientists. The interest in entering externally produced mutants into TJL has been so great, however, that a review panel has been established to select those to be included. During 1995,

100 strains were added to the Induced Mutant Resource. TJL recognizes that this constitutes only a small proportion of these rapidly proliferating strains, and it is actively soliciting funds with which to expand the IMR.

Financially, the various units have some interdependence, because proceeds from the sale of animals by Animal Resources help to offset the costs of less financially viable resources (the philosophy of the laboratory is to provide mice to researchers as close to cost as possible). Although the distinction among these various resources is transparent to the users of TJL, the division of responsibility provides more focused management and encourages long-range planning.

TJL believes that a major reason for its continuing success is that each resource is supervised by a scientist with special expertise in the specific area; quality control is a major part of the supervisor's responsibility. The scientist supervises a manager who is responsible for day-to-day activities.

Several features of the TJL structure are important to the users of the resource. First, all mice that are obtained from the facility will be of known health status and genetic quality. Any mouse stock acquired by TJL is rederived by cesarean section to eliminate the burden of infectious agents that might interfere with research, and the mutation is established on an inbred line. This importation policy reduces intercurrent disease, reduces mortality and morbidity, and enhances reproductive efficiency while decreasing the costs of monitoring for disease. The second important feature of TJL mice is that the strain will be genetically defined before it is released for use. This ensures that individuals who obtain mice will continue to receive genetically defined animals.

Preservation

A major function of TJL is the preservation of murine germ plasm via embryo freezing. If no orders are received for a particular stock for six months, the stock is usually taken out of production and maintained via cryopreservation. The laboratory also conducts research to develop additional or better methods for preserving germ plasm (e.g., sperm freezing, improving reproductive technologies).

Derivation

TJL will, on request, rederive mice, develop congenic strains by customized breeding, or maintain stocks of animals for individual scientists. Charges for these services reflect the costs involved.

Surgery

On request, TJL will provide mice that have been surgically prepared for example, by hysterectomy, vasectomy, adrenalectomy. Charges for these services reflect the costs involved.

Bioinformatics

As part of its goal to educate scientists in research, TJL issues price lists, lists of stocks with genetic information, a quarterly newsletter, data sheets on individual strains, and special newsletters devoted to specific topics. The laboratory also publishes a handbook (updated every five years) and has a World Wide Web site where most materials are available electronically. In addition, TJL is the location of the Mouse Genome Database, which provides genetic mapping and descriptive information to the worldwide scientific community.

OWNERSHIP AND ACCESS ISSUES

Ownership of mice sent to TJL is transferred to the laboratory as a condition of entry into TJL. Contributors may not attach any reach-through rights to subsequent production and distribution of stocks by recipient scientists. Mice are not accepted with a condition that they must be licensed to individual academic scientists. If contributors require licensing agreements to cover for-profit use, TJL will place a label on shipping containers for mice from these strains, making the recipient aware that a license agreement is required if the mice or research is to be used for commercial purposes. TJL does not assume any responsibility for enforcing licensing agreements. If the originator of a strain requests royalties, he or she must bear the costs of rederiving the strain, putting it on an inbred background, and cryopreservation.

One of the reasons this approach works at TJL is that individuals have an incentive to contribute their genetically modified animals to the laboratory because is assumes responsibility for the distribution of animals; gives credit to the contributors in all TJL publications, including a reference to the investigator's work; and sees that the animals are shared with fellow scientists.

The increasing ties between individuals or academia and the pharmaceutical and biotechnology industries may raise some issues that would complicate this approach to ownership. In certain cases, deposits of new stocks are being severely delayed by intellectual property concerns. Occasionally these delays are attributable to investigators, but the primary problem has been with the technical transfer offices at universities hoping to benefit financially from licensing or royalties. In fact, to date, there has been little monetary return from mice or other materials. This delay at the university level has also penalized investigators who want to share their mice and get replication or extension of results, forge collaborations, and so forth, because the investigators are then forced to use their technician and laboratory time to produce the mice for sharing with colleagues. A better understanding of university policies and what works or does not work to encourage and enable sharing of resources was identified as a major gap in the system.

Although most problems to date have been resolved satisfactorily, Jackson Laboratory administrators are nevertheless working on approaches to codify a research exemption in patent law, so that even if biologic materials are restricted or licensed, they could still be used for research purposes.

Jackson Laboratory mice are available to any scientist, regardless of employer, who wants to use them for research, with the restrictions only that they not be bred for redistribution or redistributed outside the recipient's institution. Scientists are, however, asked to come back to TJL for new breeders after 10 generations or to put their own laboratory registration code on the strain so it is no longer perceived as a JAX mouse. These restrictions serve principally to protect the genetic purity and pedigree of the animals. No mice are distributed to non-research institutions without a guaranty that there is a veterinarian to take care of the mice.

COST ISSUES

As a research institute, TJL derives considerable grant support from federal and private nonprofit agencies. This goes in large measure to the portions of the laboratory that are conducting basic genetic research, but even those units focused primarily on providing mice and related services to other scientists benefit from funding from NIH, NSF, the Howard Hughes Medical Institute, and a number of voluntary health organizations. The last of these are organizations such as the American Cancer Society, the Cystic Fibrosis Society, and the Multiple Sclerosis Society, which typically contribute amounts in the $10,000–$25,000 range for maintenance and production of mouse strains that are important to research on their own diseases. Because of this external support, the costs of mice to outside purchasers are less than they would be if all costs had to be recovered through sales of mice. TJL does distribute mice for a fee, so users pay part of the cost. Because of the grant support however, TJL can provide mice from the small specialized resources at a lower cost than would be necessary to make them self supporting. Animal Resources, which distributes high-demand strains such as C57BL/6J and other standard inbred strains, also helps cover the cost of more specialized strains so that these strains do not have to provide full cost recovery for maintenance. This maintenance and distribution of special stocks at less than full cost may not be possible in the future, as federal money becomes tighter and Animal Resources is expected to cover more of the laboratory's research costs.

Like ATCC, The Jackson Laboratory puts institutional dollars into capital investments such as buildings, renovations, and equipment, as well as nonmonetary costs involved in providing resources for others. Muriel Davisson told the workshop that TJL is spending an incredible amount of institutional time negotiating agreements to obtain specific scientifically valuable mouse strains, despite the fact that these strains generate very little monetary return, either to TJL or to the people who contribute the mice. In-house scientists also personally provide a great deal of information about the resources that they share. Even though there is a technical support crew of two, which soon will be increased to three, the scientists themselves spend a considerable amount of time with customers and prospective customers. Finally, because it is the culture at the laboratory to share anything once it has been published, TJL scientists often find their own competitively funded research programs compromised by sharing information with potential competitors.

OTHER ISSUES AND PROBLEMS

Problems that TJL has encountered include the fact that very few mouse strains are commercially viable. This has led other suppliers to develop their own mouse stocks of the most favored strains, which undermines the financial vitality of TJL. With the new importations into the Induced Mutant Resource, there are typically about 2,000 strains on campus, which is probably 100 times more than most commercial breeders would distribute. Without this additional overhead, commercial concerns do not find it difficult to undercut TJL prices on the popular strains.

Licensing requests, particularly by the contributor's institution, delay the release of new and interesting strains and add to the costs of the process. This aspect is enhanced by the increasing alliances between nonprofit institutions and for-profit biotechnology firms.

Initiation of the Induced Mutant Resource, and especially the attempt to open it to the maintenance of "knockout" mice developed elsewhere, have made it plain that a very large infusion of funds and personnel will be required if the TJL collection is ever to approach the status of a comprehensive national repository.

5

The Washington Regional Primate Research Center

The seven regional primate research centers were established by congressional mandate during the 1960s and are now funded by the National Institutes of Health (NIH) through the National Center for Research Resources (NCRR). The primate centers are distributed throughout the United States, and together they maintain more than 18,000 nonhuman primates representing 32 species. The following objectives were specified by Congress:

1. To develop nonhuman primate models for basic and clinical research and to examine the underlying mechanisms and processes of human health problems and diseases.
2. To pursue basic and applied biomedical nonhuman primate research directed toward solving human health and social problems.
3. To establish a resource for scientists from many disciplines who are trained in the use of primates and who maintain both the continuity and the high quality of scientific research.
4. To develop improved breeding practices that more adequately meet the overall research demands of the centers for high-quality, disease-free primates.
5. To continue efforts to preserve primate species threatened with extinction.
6. To provide opportunities for research experience to graduate students; postdoctoral fellows; visiting scientists; faculty members; and medical, dental, and veterinary students.
7. To identify and develop nonhuman primate models of human diseases.
8. To develop new methods and equipment for primate studies.

9. To study natural diseases of primates and techniques of importation conditioning, housing, and management, which improve the well-being and suitability of the research primate.

10. To supply biological specimens to biomedical investigators.

11. To disseminate findings of center-supported studies to the biomedical research community.

The primate centers have 190 core scientists who receive part or all of their salary and research support through the primate center grant. The core scientists, in turn, work with 924 collaborators, affiliates, or visiting scientists and have 276 graduate students. During 1994, they produced 1,200 scientific publications and books, and more than 450 are in press. These scientists are assisted by more than 1,000 support staff at the primate centers. All but one of the primate centers is directly affiliated with a university, but most of them are not located on the main campus of the affiliated institution.

When the primate center program began, each new center had an identifiable focus, which often was linked to a particular species of primate. As the centers have matured, there is increasing overlap among them in regard to the focus of their research. Despite this merging of some activities, each primate center still maintains some of its original orientation. Perhaps the major influences on the scientific programs of a primate center are the research interests of the director and core faculty, the research strengths of the institution, and the availability of funding for particular types of research.

FACILITIES AND PROGRAMS

The Washington Regional Primate Research Center (WRPRC) is an integral part of the University of Washington research community. The central campus facility is part of the Warren G. Magnuson Health Sciences Center in Seattle, which houses the schools of medicine, dentistry, nursing and public health; several other research centers; and the university medical center. The 45,000-square-foot, three-story building, dedicated in 1964, is designed specifically for primate housing and research; it houses about 500 primates: baboons (*Papio papio*), and monkeys, primarily pigtailed and cynomologus macaques (*Macaca nemestrina* and *Macaca fascicularis*). In addition to conventional laboratories and associated facilities for animal housing, cage washing, food preparation, and veterinary care, the building contains fully equipped surgical and radiological suites for experimental and clinical use and an automated quantitative microscopy system. Additional features are facilities for covertly observing and recording the behaviors and social interactions of groups of nonhuman primates, as well as certified biological safety level 3

(BSL-3) facilities for studying hazardous viruses and other pathogens. The WRPRC staff is composed of eight doctoral-level researchers. They are joined by approximately 300 other scientists and graduate, medical, dental, and veterinary students in a wide array of research projects. In recent years the center's principal research efforts have focused on neuroscience, cardiovascular physiology and pathology, hemorrhagic shock, complications of metastatic cancer, and viral diseases, including AIDS. Like the other centers, WRPRC has always supported the research and development of animal housing facilities and of breeding, rearing, and management techniques that maximize the health and well-being of laboratory primates.

WRPRC operates several additional facilities. The Infant Primate Research Laboratory, located in the health sciences complex at the Center on Human Development and Disabilities provides lab space and housing for 125 animals and a 24-hour-a-day intensive care unit for low birth weight animals, animals rejected by their mothers, or primate infants assigned to studies of developmental problems such as fetal alcohol syndrome or respiratory distress syndrome. The Primate Field Station is located on the grounds of Eastern State Hospital at Medical Lake, Washington. Its principal function is breeding animals for research (about 350 annually). A small program in southern Russia maintains a breeding colony of pigtailed macaques (*Macaca nemestrina*), and a major breeding and research program at Tinjil Island, Indonesia, produces *Macaca fascicularis*. The latter is a cooperative program of the WRPRC, Bowman Gray School of Medicine, and Institut Pertanian Bogor, Indonesia. This new program is designed to produce healthy research animals in a free-ranging natural habitat at relatively low costs. It provides scientific training in an underdeveloped country as well as helping the country benefit from its natural resources.

The Primate Information Center of the WRPRC maintains three national information systems. One of these is a very comprehensive bibliographic database on nonhuman primates (35,000 post-1984 references). *Current Primate References* is a monthly bulletin that lists new publications on primates from all around the world. The center also publishes a number of specialized kinds of bibliographies, particularly the topical bibliography, and has a very well developed primate database that can be leased by investigators with special needs.

The Primate Supply Information Clearinghouse was set up in 1977 specifically to provide information for the sharing of primates; it not only serves primate centers but is a nationwide effort to facilitate efficient use of nonhuman primates by collecting and listing offers from laboratories with available primates and requests from laboratories that need specific primates. Its primary purpose was conservation; because of the inability to exchange this kind of information in the past, many primates were being euthanized when they could have simply been used in other research programs. In recent years the database has expanded to include information on available facilities and services and on regulatory requirements. This information is disseminated through a series of publications, which include weekly and annual bulletins. The Primate Supply Information Clearinghouse has been very successful in disseminating information about the availability of primates for all kinds of primate-using facilities including primate centers, zoos, biotech companies, government agencies, and others. Any facility licensed with the U.S. Department of Agriculture (USDA) can take advantage of the service.

The Tissue Distribution Program is a spin-off of the clearinghouse. This program maintains a listing of scientists' needs and provides fresh, fixed, or frozen specimens prepared in a variety of ways to laboratories throughout the world. About 50 investigators are currently using this service, most of whom are from academic research institutions; about 3,000 tissues harvested from roughly 200 animals will be distributed this year.

OWNERSHIP AND ACCESS ISSUES

Although tissue specimens distributed through the Tissue Distribution Program become the property of the recipient, all living WRPRC monkeys are owned by the primate center regardless of their status as research subjects. Scientists with peer-reviewed, funded support, who have protocols approved by the University of Washington's Institutional Animal Care and Use Committee (IACUC), submit requests to the WRPRC's Research Review Committee to utilize monkeys. When the Research Review Committee approves the study, the scientist pays an acquisition fee and per diem for the duration of the study, but the ownership of the monkey is retained by WRPRC. This does not preclude terminal studies, nor does it necessarily prevent scientists from continuing their studies if they relocate to other institutions. Although typically the research is conducted within the primate center itself, this is not required, and significant parts of the research may be carried out elsewhere.

The current review process provides extensive oversight of proposals, but it also creates at least two types of difficulties. First, the Freedom of Information Act of the State of Washington requires that all IACUC meetings be open to the general public, and animal rights activists have used this mechanism to harass individual scientists and members of the WRPRC faculty or staff. Second, although the strength of the Washington Regional Primate Research Center is in no small measure due to the strong scientific leadership it received from prior center directors Orville Smith and Douglas Bowden, and the strong research faculty at the University of Washington, the secondary review by the WRPRC Research Review Committee creates the potential for a real (or perceived) conflict of interest by members of the committee who might be reviewing competing scientists' proposals.

For extramural scientists whose interests are similar to those of a core staff scientist at WRPRC, the easier route to access is often a collaboration, although in recent years even this arrangement has resulted in some intellectual property problems of the sort discussed in the previous case studies. William Morton pointed to his own work with a monkey AIDS model using a variant of human immunodeficiency virus (HIV) known as HIV-2287. The original HIV-2 strain or isolate was brought back from France years ago by an individual from what was then called Genetic Systems, a small biotech company in Seattle. The company evaluated a number of HIV-2 prototypes in vitro and finally chose several to study in vivo. WRPRC and Genetic Systems formed a collaborative arrangement and inoculated some of these strains into pigtail monkeys. After several years of whole blood transfusions to identify more virulent mutants, they defined a variant of HIV-2 they named HIV-2287, which can cause acute infection, CD4 cell decline within two to six weeks, and ultimately full-blown AIDS syndrome within six to nine months.

Now, however, Genetic Systems no longer exists, having been acquired by Bristol Myers. Bristol Myers would like complete ownership of HIV-2287 and has drawn up and forwarded an agreement saying that Bristol Myers owns HIV-2287, that it has the right to tell WRPRC scientists when they can or cannot use this variant, and when they can publish, and that the company may invoke confidentiality about any or all communications concerning the strain. Dr. Morton believes he cannot sign such a document. Instead, WRPRC scientists will have to try to rederive and reisolate another strain to develop their own titered stock rather than using the HIV-2287 stock, which Bristol Myers now claims as its own. This will resolve the disagreement, but it will replace more productive research.

COST ISSUES

Each regional primate research center operates under grant funds obtained from NCRR, and its program is reviewed every five years. In 1993, the seven regional primate research centers received $40.8 million (57 percent) of the $72.2 million budget of the Comparative Medicine Program of NCRR. In comparison, the Laboratory Animal Sciences Program, which support grants for research, other research resources, and training, received $22 million (30 percent) of the Comparative Medicine budget.

WRPRC received $6.7 million in core funding from NCRR in 1995, allocated to four major categories: Basic Research ($1 million), AIDS-related Research ($1.6 million), Basic Services ($2.4 million), and AIDS-Related

Services ($1.7 million). Additional sources of income are the State of Washington, which provides about $600,000 per year in the form of support for faculty salaries and indirect cost reimbursements, outside grant support for core staff and research affiliates ($22.5 million), and charges to users for animals and services ($3 million).

As noted in the previous section, investigators are charged an acquisition fee for animals and a per diem charge for the duration of the study even though WRPRC maintains ownership. Medical and surgical procedures required by experimental protocol or clinical care are also billed to the investigator, as are clinical laboratory tests. Investigators receiving specimens through the Tissue Distribution Program are charged a fee (in this case ownership transfers with the specimen). In keeping with the WRPRC philosophy that promoting basic research should be one of its primary missions, noncommercial users are given a substantial discount on all of these charges (20–50 percent).

Primate centers have high maintenance costs because nonhuman primates are extremely destructive and require high levels of containment. The costs of acquiring and maintaining the health of animals from these primate centers are significant, but the health status and welfare of the animals that they produce are certainly greater than could be obtained by trapping feral animals. Moreover, many countries no longer permit trapping or sale of monkeys.

The increasing costs associated with raising the animals are forcing reevaluation of where the breeding colonies are located. To reduce the costs of maintaining breeding colonies, the old Medical Lake facility 300 miles from Seattle will be closed in the coming years. A new, smaller, and more efficient facility at American Lake will be constructed; this facility will be shared with the Oregon Regional Primate Research Center. In addition, some breeding animals will be moved to the Tulane Regional Primate Research Center in Louisiana, where they can be reared in outdoor corrals far less expensively, at perhaps one-tenth the cost of Medical Lake. The Indonesian facility, where the colony ranges freely about the small island of Tinjil and lives off the land, promises to be even less expensive.

OTHER ISSUES AND PROBLEMS

The plans to breed healthy animals in a setting more analogous to their native habitat (such as the facility in Indonesia) may be a partial solution to the costs, This solution is, however, totally dependent on the availability of transportation for those animals. At present, virtually all international commercial airlines, under pressure from animal rights activists, have refused to transport nonhuman primates. To a much greater degree than with other laboratory animals, the transportation and handling of nonhuman primates may be associated with significant or potential biohazards such as Herpes B virus, filovirus, shigella, and others. Thus, even within the United States, transportation needs are the downside of any consolidated breeding plan.

The costs of operating the Primate Center Program are also adversely affected by increasing regulatory activity by the USDA and the Public Health Service via the Office for Protection Against Research Risks (OPRR). It is not unusual for different federal agencies to have different requirements for the same species of animals. Regulations have been imposed requiring provisions for the psychological well-being of nonhuman primates and specifying precise cage sizes. Cages for monkeys commonly cost $8,000–$10,000 each. Many of these regulations have more to do with the perception of the individuals (many of whom were nonscientists) who proposed them than with scientific data. In addition, accrediting organizations such as the American Association for the Accreditation of Laboratory Animal Care (AAALAC), as well as federal agencies, are requiring increased recordkeeping and other types of documentation. All of these regulatory requirements are coming at a time when there is less and less core support from federal funding for the infrastructure for key laboratories and administrative personnel—a role that financially strapped academic institutions facing similar constraints are unable to assume.

The political activities of animal rights organizations create major costs for a primate facility because nonhuman primates are typically perceived by the public as more sentient animals with a special bond with humans. By focusing on nonhuman primates, animal rights organization generate sympathy for animal rights, in general, and also often generate funds for the animal rights cause. The costs of animal rights activities to an institution take several forms and include

1. the costs of security to protect the safety of employees and the investigators' research;
2. the costs of security to prevent vandalism by committed animal rights terrorist groups;
3. the costs of litigation instigated as a form of harassment by animal rights activists; and
4. the costs of staff time to investigate and respond to allegations by animal rights activists of violations of animal welfare regulations. Federal regulations require that all allegations of improper care of animals reported to internal (IACUC) and external (OPRR and USDA) oversight bodies be investigated. Therefore, even frivolous complaints must be fully investigated and documented, and the findings must be reported to the appropriate agencies.

6

The Macromolecular Crystallography Resource at the Cornell High-Energy Synchrotron Source

USER FACILITIES FOR PROTEIN CRYSTALLOGRAPHY AT SYNCHROTRONS

Synchrotron x-ray sources, used extensively for diffraction studies of biomolecular structures, are an example of a mature set of shared instruments and facilities. At the present time there are eight such x-ray sources worldwide with significant capabilities for structural biology research, and three additional installations will soon to be operational or under construction. Synchrotron radiation sources are expensive and can be constructed only through the cooperation of a large research community. Typical users of synchrotron radiation sources include physicists, materials scientists, biologists, chemists, and others. The larger community must support the construction of the synchrotron storage rings (which cost several hundred million dollars), while smaller groups of researchers band together in order to instrument individual beam lines for specific types of experiments. The cost of an individual beam line is several million dollars for construction plus ongoing operating costs. Individual scientists who use these resources share x-rays, instrumentation, and some software. These facilities are a rich source of information about what works in shared facilities and where there are problem areas or bottlenecks.

The past decade has witnessed a dramatic increase in capabilities for determining the three-dimensional structures of macromolecules. New structures currently appear in high-impact journals such as *Science, Nature,* and *Cell* at rates approaching one per week and have profoundly affected every area of the biological sciences. Macromolecular structures produced by x-ray crystallography and nuclear magnetic resonance (NMR) spectroscopy are used to understand the structural basis of protein function, resulting in applications to fields such as drug discovery and protein engineering. Synchrotron radiation

THE MacCHESS RESEARCH RESOURCE

The Macromolecular Crystallography Resource at the Cornell High Energy Synchrotron Source (MacCHESS) provides support for the collection and analysis of x-ray diffraction data from crystals of biological macromolecules using synchrotron radiation. The overall goal of the MacCHESS research resource is to ensure that world class research in structural biology is performed at CHESS. This goal is accomplished by providing specialized equipment for macromolecular crystallography as well as trained support staff to assist outside users. The MacCHESS staff of two scientists, three technicians, a computer programmer, a machinist, and a secretary has established an active research program designed to advance the frontiers of synchrotron radiation research and structural biology. MacCHESS receives its major source of funding from the Biomedical Research Resource Program of the National Center for Research Resources at the National Institutes of Health (NIH).

MacCHESS is dependent on the continued operation of the CHESS laboratory, which is responsible for delivering synchrotron radiation to the experimental hutch. CHESS is funded by the National Science Foundation (NSF) to provide synchrotron radiation for a wide variety of experiments. Structural biology accounts for about 35–40 percent of the total experiments performed at CHESS. CHESS, in turn, is dependent on the operation of the CESR (the Cornell Electron-Positron Storage Ring). CESR is maintained by the Laboratory for Nuclear Studies for use in particle physics experiments and is funded by the NSF.

Core Research Projects

Core research projects are performed by MacCHESS faculty and staff and are intended to advance the capabilities of the research resource. Core research projects provide the driving force for new developments such as advanced x-ray detectors, cryo-crystallography apparatus, new beam line optics, new x-ray instrumentation, and new data analysis software. Current core research projects include structural analysis of various targets for drug design, elucidation of enzyme mechanisms and protein engineering.

Collaborative Research Projects

Collaborative research projects are intended to extend new developments to a broader research community. Collaborators obtain early access to new instruments, techniques and methods and provide additional impetus for their development and refinement. For example, collaborative research was used as a mechanism for testing new x-ray detectors and the multiple wavelength anomolous diffraction (MAD) phasing instrumentation.

User Research Projects (Service)

Mature methods are made available to outside scientists using the facility on a competitive basis. In 1995, MacCHESS users performed experiments that included preliminary crystallographic analysis, high-resolution data collection, data collection for large unit cells, multiple isomorphous replacement (MIR) structure determination, molecular replacement structure and MAD phasing experiments. More than 200 scientists from 45 laboratories used CHESS facilities for macromolecular crystallography experiments during 1994. This work resulted in dozens of scientific publications and presentations at meetings.

Training and Dissemination

Workshops and Symposia

As a user facility, MacCHESS provides visiting scientists with on-site training for all aspects of macromolecular crystallography including crystal freezing, experimental design, operation of station bench cameras, use of both image plate scanners and charge-coupled device (CCD)-based x-ray detectors, and evaluation and processing of data using various processing programs. Each year, CHESS organizes a users' meeting and workshop. The users' meeting features reports of research and development activities by both the local staff and outside users. The topic of the workshop relates to macromolecular crystallography about every other year.

Training Videos

MacCHESS has produced the first of a series of training videos on macromolecular crystallography using equipment provided by the Keck Laboratory for Molecular Structure at Cornell University. The first training video focused on cryo-crystallography and has been distributed to more than 400 scientists worldwide. Other videos covering various aspects of macromolecular crystallography and synchrotron radiation are planned.

CHESS Newsletter

Each year, CHESS publishes a newsletter that highlights the productivity and capabilities of the CHESS laboratory. In recent years, nearly half of the contributions have been in the area of structural biology. Future newsletters are planned to keep the community informed about ongoing and planned CHESS activities.

MacCHESS World Wide Web (WWW) Home Page

MacCHESS has established a WWW home page with which users can keep up with the latest developments in instrumentation, software, progress, and opportunities from MacCHESS. From a separate CHESS home page, users can learn about new CHESS developments and obtain beam time application forms. The WWW page has already proved to be an effective way for users to remain informed about MacCHESS in the time between CHESS newsletters.

OWNERSHIP AND ACCESS ISSUES

Synchrotron sources originally served primarily as research facilities for a select group of participating scientists who were developing the methodology and the technology. Now, however, there is an strongly increasing user demand for access to these facilities, particularly by the crystallography community. Access to CHESS by outside investigators is through competitive proposals based on the peer review process. Four types of proposals are available: Program Proposals, Standard Proposals, Express Mode Proposals, and Feasibility Studies. Deadlines for Standard Proposals (requesting a block or blocks of time for a single experiment or structure, with approval good for two years) and Program Proposals (for a series of linked experiments or related structures over a four-year period) are announced about every six months. The proposals are sent to external reviewers for evaluation, and a final priority score is assigned by a proposal evaluation committee comprised of scientists representing all major areas of synchrotron radiation research. Access to CHESS is based on scheduling requirements and the final priority score.

Express Mode Proposals were implemented to address the need of macromolecular crystallographers for rapid, short-term access to synchrotron radiation. A portion of the total beam time is set aside for these purposes based on the level of demand but is restricted to limit adverse effects on other types of proposals. Express Mode Proposals are normally limited to a maximum of 48 hours of beam time and should not involve hazardous materials. Express Mode Proposals are evaluated by a three-person committee, and beam time is allocated based on the committee's evaluation.

Feasibility Studies are short-term access (up to four days) proposals that provide greater flexibility than Express Mode Proposals. The proposals are evaluated by a separate committee, and beam time is allocated based on the committee's evaluation. Feasibility Studies may involve hazardous materials and therefore may require approval from the Safety Committee.

The procedure works satisfactorily for most users, and about 40–50 percent of the good projects gain access to synchrotron time. Access is currently reviewed independently of grant support for the projects involved, which potentially creates a chicken or egg dilemma. However, since synchrotron time is even more limited than grant support, the issue of awarding time to an unfunded project appears to be rarely, if ever, faced.

Some of the most active scientists using synchrotron time are peripatetic wanderers who submit multiple applications to multiple facilities and use time wherever and whenever it can be found. To date, the system appears to have proved itself to be reasonably capable of dealing with this bit of redundancy. However, peer review of the same proposals for beam time by committees at several sites will increase significantly as the number of stations for protein crystallography studies in the United States approximately doubles in the next few years (up from the current 9 to about 20). Coordination of proposal review and scheduling among all facilities would be a significant logistical task, but facility directors should begin to explore mechanisms of coordination along with simplified review and scheduling algorithms that could reduce the administrative burden significantly.

Rapidly increasing demand from nonspecialists (i.e., biologists or other scientists with an interesting molecule but no experience with a synchrotron or maybe even with crystallography) has underlined a problem facing all user facilities. There is a need to balance core research, which keeps the staff enthused, with collaborations and service to outside investigators. Despite the addition of two new beam lines, MacCHESS still has the same staff and budget as when it had only one, and a lengthy backlog of approved projects waiting for beam time has developed. The director is actively seeking new sources of funding for core research, in order to focus more of the existing staff's effort on user support.

Ownership issues are perhaps less complex at MacCHESS than with some of the other case studies in this report. In the typical study, an investigator comes with his or her crystals and leaves with all the data. That said, there are also collaborations involving core staff and outside investigators, and there is no hard-and-fast rule about authorship at MacCHESS. The director's view is that purely technical assistance deserves an acknowledgment in any subsequent papers (and authors who overlook that are quickly notified), but that the basis of coauthorship negotiations should be contribution to the interpretation of data (as opposed to simply enabling data collection).

COST ISSUES

As noted above, MacCHESS is supported by a grant from the National Center for Research Resources (NCRR) at NIH. That funding comes to approximately $1 million annually. This apparently straightforward arrangement is complicated by the fact that MacCHESS is dependent on CHESS, which has an annual budget of $2 million funded by NSF. CHESS in turn is dependent upon the operation of the half-mile-circumference CESR, which has annual operating costs of $15–20 million, also provided by NSF.

Underutilization of the facilities because of insufficient funds to keep the synchrotron running throughout the year, or because of the competing needs of biological users and high-energy physics users, has been a significant problem for the other four U.S. synchrotron sources, all of which are operated by the Department of Energy (DOE). The DOE scientific facilities initiative of FY 1996 provided these facilities with an increase in operational funding to ensure full-time synchrotron operation. Cornell is the only synchrotron source funded by NSF and, thus has been affected only indirectly (changing demand for beam time) by these changes in DOE funding, but the dependence of the structural biology community upon support for a very high budget physics program is an obvious hazard in an era of tight money.

At the level of MacCHESS, it is important to note that the basis of NCRR support is cutting-edge methodological research by core staff, rather than service to structural biologists from other institutions. This has not been a problem to date, but it does allow the possibility of success (in attracting users) putting a shared resource out of business.

User fees generally are not assessed against the grants of outside investigators, although they are charged the costs of consumable supplies (x-ray film, etc.). Commercial users who insist that their work is proprietary are however charged for beam time (including the necessary services of staff) at the rate of about $800 per day. This charge seems unlikely to offset the full cost but may cover incremental costs. Proposals from industry are at somewhat of a disadvantage in the review process leading to beam time scheduling, because industry proposals generally deal with protein structures that are already known (e.g., human immunodeficiency virus (HIV) protease). At least at present, the prospect of breaking entirely new ground seems to weigh heavily in the review committee's decisions. Perhaps for this reason and because of the need to meet developmental timetables, a group called the Industrial Macromolecular Crystallography Association has raised enough money to build two beam lines at the advanced photon source about to open at Argonne National Laboratory. In return for "their own" beam lines, they have promised to give 25 percent of beam time to independent outside users. A similar arrangement might be a solution to both the handicaps faced by industry submissions in the review process at MacCHESS and the increasing demand for services of the staff by inexperienced users.

OTHER ISSUES AND PROBLEMS

In the past, the effectiveness of synchrotron sources was often compromised for rather picayune or cost-ineffective reasons. Local infrastructure for biological experiments has often not been topflight at many synchrotron sources. This needs to be watched and corrected since the costs are often trivial compared with the cost of otherwise wasted synchrotron time. The current situation is reported to be largely satisfactory.

Another area in which progress is needed is stronger local software support at the facilities. This will allow preliminary data analysis to be carried out almost in synchrony with the gathering of experimental data so that some problems can be caught early and corrected and the sheer volume of unprocessed data that has to be carried off-site and kept intact can be reduced to an acceptable level.

A final problem that threatens to limit the utility of multiuser synchrotron sources is the requirement to travel to the site. Traveling to such remote facilities is an experience outside the culture of most biomedical or biological researchers. This will have to change as unique, expensive facilities become the norm in other areas such as very high resolution magnetic resonance imaging, accelerator mass spectrometry, and very large scale DNA sequencing to name just a few. A key issue for the design and operation of such facilities is whether the experimental users must come to the site or whether just their samples can be sent. Clearly, the more the latter mode of operation can be adopted, the less disruptive and the more effective shared resources will be. There seems to be no reason why, for the many routine types of applications, remote access will not suffice, and the sorts of remote monitoring that have become common in medical practice could easily be adopted to handle the vast majority of experimental situations. What must be accepted is the local cost of supporting significantly increased staffing at these facilities, whose role would be mainly—if not entirely—the support and service of external users. The danger in this model is possible stagnation in the continued development of novel capabilities for new science. Attention must be paid to a balance between these competing needs (a competition likely to be exacerbated in an era of restrained funding).

7

The Human Genome Center: Lawrence Livermore National Laboratory

The last of the six case studies examined is neither a repository of scientific community property, like the American Type Culture Collection, nor a center for visiting scientists, like the Cornell synchrotron. The Human Genome Center at Lawrence Livermore National Laboratory in fact is not technically a shared resource at all, but a federally owned, contractor-operated research and development laboratory that has become, by default, a supplier of valuable materials and information to the international scientific community without specific funding to do so. The resources shared by Livermore fall into several categories described below.

LAWRENCE LIVERMORE NATIONAL LABORATORY

Lawrence Livermore National Laboratory (LLNL) is one of several Department of Energy (DOE) national laboratories. Located about 50 miles southeast of San Francisco in the Livermore Valley, it is operated for DOE by the University of California. Founded in 1952 by E.O. Lawrence as a second nuclear weapons laboratory (joining Los Alamos), LLNL now applies its skills in high-performance computing, advanced engineering, and management of large research and development projects, to a broad range of the nation's technological challenges. The laboratory has 12 scientific or engineering directorates; the Biology and Biotechnology Research Program, which houses the Human Genome Center, is one of those 12 directorates.

When it began in 1963 as the Biology and Environmental Research Program, its initial task was to study the dose to man of isotopes in the environment as a result of fallout from weapons testing, and the immunologic and genetic consequences of radiation. By 1992, the program was doing little radiation work but was heavily involved in studying the mechanisms of genetic damage from toxic chemicals and disease. A reorganization yielded the Biology and Biotechnology Research Directorate, with an increased emphasis on biotechnology and structural biology, while environmental studies joined similar efforts in the Environmental Programs Directorate.

The Biology and Biotechnology Research Program has an annual funding level of approximately $30 million, primarily from peer-reviewed research grants. About half of that funding is through Department of Energy grants; the other half is from various sources, including the National Institutes of Health (NIH) and other federal agencies, as well as industry. Its activities fall under four programs:

1. **Health Effects** assesses exposure to toxic agents, carcinogens, and mutagens for a wide variety of sources; also studies DNA repair, the genetics of cancer susceptibility, and biodosimetry.

2. **Health Care** applies Livermore-developed technology to disease detection and treatment.

3. **Structure-Function Analysis** investigates proteins and other molecules responsible for maintaining the integrity of the human genome.

4. **Genomics** develops recombinant DNA clones, DNA mapping and sequencing techniques, and instrumentation and informatics tools to characterize the genes of microorganisms, animals, and humans. The LLNL Human Genome Center is the focus of this activity.

HUMAN GENOME CENTER

DOE has established genome centers at three sites, Los Alamos National Laboratory (LANL), Lawrence Berkeley National Laboratory (LBNL), and Lawrence Livermore National Laboratory (LLNL). Together with 18 multidisciplinary Genome Science and Technology Centers supported by the NIH National Center of Human Genome Research, they form the backbone of the Human Genome Project. The Human Genome Center at LLNL was established in 1990 as an outgrowth of ongoing work on DNA repair genes, specifically on chromosome 19. A multidisciplinary team of chemists, biologists, physicists, mathematicians, engineers, and computer scientists, the center is organized into four broad areas: Resources, Physical Mapping, DNA Sequencing, and Enabling Technologies. Each area consists of multiple projects led by a principal investigator. Together they labor at three tasks—creating biological resources useful for genomic research, developing instrumentation and informatics for genome research, and locating genes.

The **Resources Group** provides cloned DNAs, synthetic DNA oligomers, DNA library production, DNA fingerprinting, restriction mapping, and fluorescence in situ hybridization (FISH) for the center, as well as maintaining all relevant data in an easily accessible database.

The **Enabling Technologies Group** provides center biologists with support in genome informatics, covering data acquisition, storage, integration, and display; with statistical and mathematical expertise critical to mapping and sequencing; and with instrumentation in such areas as fluorescence, electrophoresis, and high- throughput sample handling.

The **Physical Mapping Group** localizes biologically interesting features (genetic markers, genes, and regions conserved between species) of the genome, with special emphasis on human chromosome 19.

The **Sequencing Group** works in three areas: (1) highly accurate "finished" sequencing of genomic DNA containing genes of interest; (2) support sequencing for other researchers, including partial and full-length cDNA sequencing and generation of sequence tagged sites (STS); and (3) sequencing of large sections of chromosome 19.

Livermore chose chromosome 19 as its initial target and has completed an integrated metric physical map of human chromosome 19 that spans 95 percent of the euchromatin. The center has also developed and applied new biochemical and mathematical approaches for constructing ordered clone maps and has sequenced several DNA repair genes in humans and rodent species. Impressive as the center's own research has been, it is the unintended consequences of such success that are most relevant to this report. The center's success has produced a demand for materials and information from other scientists that is far beyond what center staff and budget can meet. The following sections describe the center's experiences in attempting to share locally developed instrumentation and technology, information, and biological materials with the wider scientific community.

Instrumentation and Technology

Instrumentation and technology is the area in which LLNL and center policies and experiences have been most conventional and most "successful." Instrumentation with commercial potential is basically kept proprietary until an invention disclosure or patent application is filed. Recent examples include a miniature diode laser-based miniflow flow system and a battery-operated portable PCR (polymerase chain reaction) device. In September 1995, a patent was awarded to the University of California for a technology developed by center scientists called "chromosome painting," which uses FISH to stain specific locations on chromosomes. The center hopes to make this technology widely available through a nonexclusive sublicensing program.

The center also develops noncommercial instruments, for example, devices that do not appear to have commercial potential because they are either modifications of something that already exists or simply because the center did not think potential demand justified a proprietary approach. Access to various instruments is given to external users on an as-available basis, even though the center is not technically a user facility, on the condition that users provide their own people and their own reagents. Center projects are always the first priority however, and outside laboratories often ask to buy a replica of the instrument instead. When this happens, such instruments are usually provided at cost or requestors are given the designs and allowed to build their own. One example is the LLNL high-speed flow sorter. One of the other national laboratories provided the funds for the center to build a replica of its high-speed flow sorter. At the same time, the center was approached by a commercial firm hoping to market the instrument and, in fact, agreed to a limited exclusive license for that instrument.

This approach is not limited to instruments of modest cost or to U.S. laboratories. A private foundation in the United Kingdom has asked LLNL for help in building a smaller version of a very large, very expensive biological accelerator mass spectrometer currently in use at Livermore.

Information

The center maintains an extensive database for chromosome 19 information and transfers data monthly to the Genome Data Base at Johns Hopkins. Many of those data are available to anyone with access to the Internet. The center home page includes a map of chromosome 19 that is updated every six months. This map does not reflect all information on hand, even at the moment of updating, partly because of lack of staff and partly because of obligations to collaborators. It is important to note that the chromosome database and map include not just data gathered by center scientists at LLNL, but data from scientists working on chromosome 19 genes all over the world. It is a big advantage, probably a necessity, for all involved to have their own data put into context in a single database and map. This is the primary incentive for these scientists to forward their data and/or clones at an early stage. On the other hand, they are not always anxious for potential competitors to have the fruits of their labor too early in the process. The Human Genome Project is dependent upon, and therefore insists upon, sharing however, and the LLNL center's general rules call for holding unconfirmed data private until verified, but for no more than six months. Some exceptions may be negotiated for proprietary or otherwise sensitive data, but in no case will the data be kept private after publication. The center will occasionally act as a matchmaker when two laboratories with data in the six-month privacy window would obviously benefit from collaboration. Some laboratories still refuse to contribute to the database under these policies, but in general the chromosome 19 community seems to have accepted the policies as necessary for the health of the field. Further details of the LLNL Human Genome Center policies can be found below in the section on ownership and access issues.

Biological Materials

Because of the labor-intensive nature of producing them, sharing genome-related materials with all who want them has proved far more difficult for LLNL, despite advances in automation. The center's attempts to cope with requests of this sort in fact constitute the major contribution of the LLNL case study to this report.

Livermore's initial involvement in the Human Genome Project was a joint effort with Los Alamos National Laboratory called the National Laboratory Gene Library Project, a task now essentially complete. All the libraries were made. They consisted of human chromosome-specific lambda libraries, both small insert and large insert, as well as human chromosome-specific cosmid libraries, and these were distributed to the scientific community after initial quality control.

The general policy followed in this distribution is that there is no bar to further distribution by the recipients of those libraries. Commercial use of individual clones is permitted, but not of the libraries. The center wanted to preclude the possibility that someone would tie up the libraries in some commercial agreement that prevents their use by the rest of the community, but people can take individual clones and commercialize them. The only quid pro quo requested of the users is acknowledgment in publications or in presentations that they received those libraries or those clones through the National Laboratory Gene Library Project, a request the center director estimates has been honored at least half the time.

The lambda libraries are probably the easiest to handle. Quality control of these libraries was performed in-house, although because of the large amount of quality control required, selected external laboratories were involved as well. In fact the center was dissatisfied with many of these laboratories and in the end had to recheck most of the work. Those clones were provided to the American Type Culture Collection (ATCC) for distribution. ATCC had received some initial funding from DOE to subsidize the cost of that distribution, which seems to be working quite well. Several thousand of those libraries have been distributed worldwide. LLNL (and Los Alamos) receive no monetary remuneration for the clones or the libraries that go to ATCC.

Cosmid libraries are a little more difficult; so Livermore and Los Alamos basically split the 24 chromosomes present in the human genome. In some cases, multiple libraries were made. Chromosome 22, for example, had three libraries made of it, and the X chromosome had two libraries made of it. In all cases, these libraries were very extensive. The plan was to make at least a fivefold-deep library—in many cases they are 10- or 15-fold deep—and eventually develop an arrayed library in microtiter dishes.

This work took a lot of effort, and the center could not make libraries fast enough for the users who wanted them. Livermore distributed about 85 or 90 of these arrayed libraries and Los Alamos about the equivalent number, peaking in 1993 when the center had two full-time staff who did nothing but worry about this distribution and the rearraying. At that point in 1994, the center realized it just could not afford this work anymore and basically told the world that the services was being discontinued. The notion was that since there were now at least 82 copies from Livermore and an equivalent number from Los Alamos at other laboratories, it was time for those laboratories to assume the burden of further sharing.

A problem arose immediately. Many of the laboratories did not want to distribute the libraries, despite the fact that the center gave them permission, because they did not have the resources (staff, money, time, etc.). The center then tried a second approach, which was to tell laboratories requesting libraries, "Okay, you send people and supplies to our laboratory; our people will show yours how to rearray the libraries, and you can take them back with you." Unfortunately, it took only four months to discover that center staff were spending just as much time doing this work as if they had done it themselves, leaving them in their present quandry about what to do next.

One of the things that the staff have already done is provide plate pools to ATCC for distribution. These are nonarrayed sets of cosmid libraries. They have also reduced the amount of work that has to be done on rearraying by changing from 96-well microtiter trays to 384-well microtiter trays, which itself was costly in terms of staff time (two people doing this for almost a year to get the libraries rearrayed and to have at least a fivefold-deep, five-replicate set of each library).

The cost of arraying and shipping 150-some-odd libraries produced by the two national laboratories, at about $5,000 each, comes to about $750,000, which does not include the cost of having to rearray or set them up again. Both LLNL and Los Alamos have decided that it has become too costly for either laboratory to consider such distribution without some additional funding. Also it is not clearly in the best interest of a research and development laboratory to do this work out of a budget that provides no funds for such service to other researchers.

Solicitation for external vendors has met with mixed success. The two center directors, with the encouragement of DOE, went to *Commerce Business Daily* and asked whether any industrial groups out there were willing to screen these libraries and distribute individual clones to end users. They included two qualifiers: (1) the group would not transfer those libraries to a third party for distribution, and (2) the libraries should not be rearrayed so as to create new clone libraries. The clones could be distributed, but the libraries could not be further distributed beyond them, in part because of ongoing collaborations on some of these libraries and some of the clones in them. The second qualifier was that any proposal for commercialization come back to the original national laboratory for permission to continue. The United Kingdom Resource Center and the German Resource Center both expressed great interest in being distributors and had no problem with the restrictions. They will be recipients of the libraries. However, only two responses were received from for-profit companies in this country, and neither was willing to meet the requirements of the qualifiers.

Chromosome 19-Specific Cosmids

The Human Genome Center at LLNL has a very high resolution metric map of cosmids that spans nearly the entire chromosome 19. The correct order of the cosmids is known, as are the location and size of the gaps, and individual genes have been located both relative to each other and by restriction fragment. Other researchers hoping to find additional genes thus have a powerful incentive to contribute their candidate clones to LLNL, where they can be located within 50 kilobases of another marker. The resulting burden on center staff is substantial, however, and in recent years the center has taken a different tack. Robotics are used to make high-density filters on which the equivalent of 38,400 clones spanning the whole of chromosome 19 can be placed systematically. The resulting library is then sent to the would-be collaborator, who does the hybridization and notes which location "lights up." LLNL then pulls the corresponding clone from the freezer, verifies the match, and sends it to the collaborator for further research. More than 2,000 such clones had been sent to collaborators through 1995.

The center aggressively pursues collaborations, with subsequent involvement in publications, in part because laboratory policy requires that user fees, if charged, must offset the entire cost. Because its staff believes this would constitute a powerful disincentive for academic scientists, the center has chosen to distribute libraries to collaborators without charge.

IMAGE Consortium cDNA Clones

The Human Genome Center at LLNL is also a central player in the Integrated Molecular Analysis of Genome Expression (IMAGE) Consortium, along with Washington University and Merck and Company. This group, whose founders included a scientist from the staff of Livermore, is dedicated to the widespread disbursement of cDNA clones. LLNL is basically the consortium's receiving center, accepting and arraying libraries from universities and industry, checking against a master array to minimize redundancy, and updating a tracking database. The clones are then sent to Washington University, and the information is transmitted to the dbEST Database at NIH. LLNL initially took on the task of distributing clones to requestors, but found that, as with chromosome 19, demand soon outstripped the center's capability, even with one staff member committed to the task full-time. Solicitation through *Commerce Business Daily* was more successful in this case however, and there are now several distribution centers, including ATCC, from which interested laboratories can obtain clones.

OWNERSHIP AND ACCESS ISSUES

The Human Genome Center has responded to the exponential growth in requests for information and material resources by developing explicit guidelines and agreement forms, the essence of which has been conveyed above, but they will be summarized here for convenience.

Information and materials developed by or relinquished to LLNL and *published* in the open literature will be made freely available to the scientific community, to the extent that the center has the resources to comply with the request. In rare cases, requests for clones or other materials provided to LLNL by third-party collaborators may be denied if those collaborators have requested an exception to the LLNL release policy. The center requests that recipients maintain the clone or probe names assigned by LLNL and that subsequent reports and publications acknowledge contribution of such data or materials by LLNL. Libraries, clones, and other reagents provided may not be distributed beyond the recipient's lab, nor may they be used for commercial purposes without explicit permission of LLNL.

Unpublished information and materials are generally available either through collaboration or after the lapse of a suitable time period, generally six months from the date the material or data are entered into the LLNL database. Collaboration with one or more members of the center staff is the preferred mode of access. Collaborators are requested to provide LLNL with something of value to it—such as clones, probes, information, joint publications. Those providing probes or primers will in turn be sent cosmids or YACs (yeast artificial chromosomes) that are positive, along with information on the contig status of these clones.

In the event of a request for collaboration within an area in which LLNL has an existing collaboration with a third party, the existing and potential collaborators will be notified and agreement to expand the existing collaboration sought. If no agreement can be reached, the original collaboration will be honored, and the six-month hold rule will govern release of data to the scientific community.

SUMMARY OF ISSUES AND PROBLEMS

The issues and problems experienced by Livermore and its long and successful history of resource sharing are informative. The biggest issue is inadequate funding to support what has become exponential growth of demands for the products of successful programs. As the programs have outgrown the capacity of the initiating institutions to support them and distribute their products, there has generally been difficulty identifying other institutions (commercial or nonprofit) willing to assume responsibility for the expanding project and to make readily available, without constraints, the materials and information, even when charging fees.

The unique funding aspects of LLNL require that it either charge a fully allocated cost for materials and information or that the same be provided free of charge, except for shipping costs. The former would be prohibitively expensive and, according to the center director, would stymie resource sharing.

Another not infrequent problem is that multiple investigators utilizing the resources of the Human Genome Center, (i.e., sharing those resources) are not willing to share with each other, even when another (competitive) investigator might be in the immediate geographic proximity.

It seems that the pattern at Livermore is to initiate a project but as it becomes successful, the laboratory is not able to keep up with the demands of sharing. The center then decides to cease distribution other than providing it on a one time basis to subdistribution centers, if and when such can be found. LLNL then will recreate the same phenomenon with a newer technology and again exceed its capacity for sharing—perhaps an unavoidable price of success, but a price that may not be in the best interests of science.

8

Conclusions and Recommendations

The foregoing case studies by no means exhaust the list of successful efforts to share biomedical data, materials and facilities with the scientific community as a whole, but the common themes that emerged in discussion of this diverse group of cases encourage the committee to believe that they are representative of the equally successful ventures not considered because of constraints on the committee's time, energy, and funding. These common themes demonstrate some of the necessary ingredients for successful resource sharing, but also surface issues or problems that require further study.

FEATURES OF SUCCESSFUL RESOURCE SHARING

Strong Scientific Leadership in Agencies and the Research Community

Essential ingredients in successful resource sharing are the leadership of program managers in government agencies who identify opportunities and support them; the leadership of senior scientists who establish the norm for the scientific community by example and commitment to sharing resources; the leadership of scientists who direct existing shared resources to provide quality services at moderate costs; and the commitment of scientific institutions such as universities and professional societies that develop policies to facilitate and enforce resource sharing. The *Arabidopsis thaliana* genome project's remarkable communal spirit and international character have made it a successful model for scientific cooperation and sharing of research resources.

This project began when program managers in government agencies, recognizing that work on mapping and sequencing the genome of *Arabidopsis* was accelerating, convened an international series of workshops of leading scientists to devise a long-range plan. The continued commitment of these senior scientists to widespread sharing of information and materials, and the peer pressure and aggressive solicitation of stocks of mutant strains to be made available through distribution centers, have contributed to the almost universal sharing of materials in this community. Similarly the strong leadership of the 22 societies that provide oversight for the American Type Culture Collection (ATCC), and the strong scientific leadership and management of The Jackson Laboratories (TJL) are strengths of these successful repositories and distributors of resources. A most remarkable example is presented by the Human Genome Center of the Lawrence Livermore National Laboratory (LLNL), which, by default, has become a major supplier of material resources to the scientific community, without being supported for this function. The extent to which it has provided the leadership and the actual materials that have permitted widespread sharing of genetic materials and information and the forging of important collaborations is remarkable. LLNL has protected the use of this important resource for the research community.

Many of the important institutions in science have an ongoing responsibility to foster a culture of sharing and to continue to advocate for policies that assist the process. Professional societies and journal editors can support sharing of resources by developing appropriate policies guiding publications and responsibilities for making data available after publication. The *Journal of Biological Chemistry*, for example, has such a policy: "Authors of papers published in the journal are obligated to honor any reasonable request by qualified investigators for unique propagative materials such as cell lines, hybridomas, and DNA clones that are described in the paper." Plans are under way to modify the phraseology to restrict the obligation to investigators who want to use the strain for noncommercial purposes and to include computer programs in the materials that have to be shared. In addition, after considerable debate, the policy was established that authors publishing crystallographic data must submit the details, coordinates, and related data to the Protein Data Bank at Brookhaven before publication. The appropriate accession number must be inserted into the manuscript; in a similar way, nucleotide sequences must be submitted to Genbank or a similar database, and the accession number must be inserted into the manuscript.

Adequate Core Funding

The committee observed that an essential ingredient for successful shared facilities or repositories was adequate funding of the core functions. In many cases there is a patchwork of funding from a number of different funding agencies, industry, and grants to support research or further development of the resource, as well as user fees. Sometimes the different streams of dollars may not be available to support the core administration and quality control necessary for resource sharing. This is inefficient and requires much effort on the part of the staff to write numerous proposals to different agencies. For example, at ATCC, decreasing core support is a cause for concern that has forced management to raise costs to purchasers to undesirable levels. The MacCHESS (Macromolecular Crystallography Resource at the Cornell High-Energy Synchrotron Source) case story is an excellent example of coordinated agency and industry support. The National Institutes of Health (NIH) is able to piggyback on the support provided by Department of Energy (DOE) and the National Science Foundation (NSF) to open these facilities for use by the biomedical community. The DOE scientific facilities initiative of FY 1996 provided these facilities with increased operational funding to ensure full-time operations and effective running. The seven regional primate research centers established by specific legislation during the 1960s and funded through the National Center for Research Resources are additional excellent examples of shared resources that have stable core funding.

Marketing and Advertising

Advertising, marketing, and general knowledge of the availability of a resource are essential to widespread access; many resources are not shared simply because their existence is not known to scientists who require them. All of the case examples studied in this workshop have a variety of mechanisms for alerting the research community about the availability and costs of their resources. From a marketing point of view, for example, ATCC has a very heterogeneous user group, supplying materials to the clinical, industrial research, university, and government markets, and it reaches these groups through a variety of printed media, electronic media, and workshops. The Jackson Laboratory provides a variety of price lists, lists of stocks with genetic information, data sheets on individual strains, newsletters, and a handbook on doing research in mice. Most of these are also available electronically. A unique resource is the Primate Information Clearinghouse set up by the Washington Regional Primate Research Center (WRPRC) in 1977. This is an international effort to list available primates and researchers desiring primates, as well as to provide literature reviews and other information such as annual reports, and regulations. The goal of this very extensive effort is to ensure that every animal is utilized to its fullest extent in research to minimize waste or needless use of animals.

Clear Guidelines About Ownership and Access

The cases reviewed at the workshop demonstrated the value of clear guidelines concerning access and ownership, although these differ depending on the resource. No single approach can accommodate the different uses or needs. Project planning should include guidelines for sharing—under what circumstances and with whom data and materials will be shared. This is an essential ingredient in preventing later misunderstandings and problems. There is increasing desire to commercialize and realize the economic benefits of biomedical research, which makes this an especially important and changing feature of shared resources. At ATCC, special collections are being developed with restricted access, and new policies have been formulated to clarify ownership at the time of deposit, with a heavy emphasis on donation to ATCC with no restrictions. In the case of *Arabidopsis*, the stock centers and databases do not permit restrictions on materials, and strong scientific leadership and peer pressure serve to make these materials and the data freely available to the research community. The Jackson Laboratory provides another example of a resource that has developed explicit policies on ownership and access, and is resisting licensing agreements or agreements that give reach-through rights to commercial entities. The Human Genome Center at LLNL similarly has developed policies to address access to information and materials it distributes in order to protect access for the rest of the research community. For example, LLNL has no bar to commercial use of individual clones but does bar commercial use of whole chromosome-specific libraries.

User Fees

One important source of funding for shared resources can be user fees. These charges help to subsidize the core operations and maintenance of those research resources that are not currently commercially viable. In addition, at both TJL and ATCC, fees from sales (mice at The Jackson Laboratory and cultures and cell lines at ATCC) help defray the costs of functions such as authentication and quality control, which are essential, if invisible, elements of first-class science.

Clear Policies for Retaining and Discarding Data and Material

There are substantial costs associated with sharing of materials and data. Policies for the disposition of materials and information that are no longer of value will be increasingly important as the body of resources that need to be shared continues to increase more rapidly than the funding available to support them. At The Jackson Laboratory, for example, if a mouse strain is not requested for six months, the strain is stored through cryopreservation, but live colonies are no longer maintained. Prioritizing which resources to support and which not to support will be increasingly important. When the growth of different induced genetic mouse strains recently outpaced the capacity of TJL to produce these for the larger research community, the laboratory established an advisory committee to decide priorities as well as seek additional funding from government agencies.

Quality Control

A critical attribute of a shared resource is that the distributed resource be what it is purported to be. Similarly, mechanisms to ensure the highest-quality research at limited-access resources such as a synchrotron are essential to their ongoing success. The Jackson Laboratory is an excellent example of intensive quality control. First, all mice obtained from the facility are of known health status and genetic quality. Any mouse released by TJL is genetically defined so that individuals who obtain mice will continue to receive genetically identical animals. Strict distribution rules protect and ensure the quality of TJL animals. Scientists are asked to return for new breeders after 10 generations and to limit distribution to their own institution. ATCC also has a long history of providing well-defined and reliable cultures to the research community.

MacCHESS, which represents a saturable resource and thus a different dimension of quality issues, has developed an excellent proposal process and peer review system to facilitate access to the synchrotron and to ensure that only the highest-quality research is conducted at the facility.

Well-Defined Policies for Function of Research and Service at the Facility

The balance between service and research by staff is a fundamental question to be considered by all centralized facilities designed to be resource centers for the scientific community. A shared resource is greatly enhanced by the presence of an excellent scientific staff that is conducting research to improve the resource and can ensure the quality of the materials. Strong scientists at the resource can also collaborate with and expand the ability of outside scientists to contribute to new knowledge. All of the case studies have strong scientific staff that conduct research to develop the resources, are critical to quality control, and also collaborate widely with outside scientists.

Sophisticated Information Retrieval and Transfer Systems

Rapid exchange of information and widespread access to data are greatly facilitated by sophisticated information retrieval and transfer systems. Rapidly evolving information systems are transforming the way research is conducted and disseminated. A decade ago, a paper that reported an extensive body of DNA sequence data was a landmark. Now such data cannot be published in scientific journals at all but are deposited in data banks. In the case of the *Arabidopsis* community, a sophisticated set of databases and links among them facilitates reaching the entire research community on an ongoing and almost instantaneous fashion. As soon as genes are sequenced in Chris Somerville's laboratory, for example, the data are sent directly to the University of Minnesota, where the initial analysis takes place. Similarly, information generated by LLNL staff and collaborators goes into the genome database funded by DOE, where the rest of the scientific community has ready access to the information.

ISSUES AND PROBLEMS

No meaningful argument can be made against the sharing of scientific resources. No convincing example exists where sharing has had preponderantly damaging or deleterious effects. Sharing almost always results in a total cost reduction, allowing existing research dollars to support a larger total research effort. Sharing has other side benefits including the rapid diffusion of new techniques or methods throughout the scientific user community and, quite often, the catalysis of scientific collaborations based directly or inadvertently on the sharing experience. The issue is, then, not whether there should be sharing, but how to optimize it. The case studies, although providing many good examples of "best practices," also provided the committee with a wealth of unresolved issues and emerging problems that any future sharing effort will have to address.

One Uniform Policy on Resource Sharing Is Not Possible

The problems of sharing resources are diverse. Solutions therefore will be similarly diverse. There are differences in the resources to be shared, the needs of stakeholders, and the distribution of resources that stakeholders command. In gathering the material for this report, the committee has dealt with the sharing of data, materials (including experimental subjects), and equipment. It is clear that the optimal procedures for sharing these three classes will differ in most cases. With data, the incremental cost of sharing or wide distribution may be negligible. Thus, sharing as broadly as possible should be the community norm. The amount of regulation or review needed to ensure standards and effectiveness in such sharing can be minimal. Successful examples of such sharing include the nucleic acid, protein sequence, and similar databases (e.g., Genbank, DNA Database of Japan, Genome Science Data Base, SwissProt, Protein Data Base, Genome Data Base), which operate as worldwide consortia with free access to all users.

Materials (or experimental subjects) fall into two classes. Some materials are renewable. Examples are clones, polymerase chain reaction (PCR) primers, strains, and most transgenic animals. Here broad sharing is to be encouraged because it is cost-effective. However, the incremental costs of sharing are significant, and mechanisms to distribute these costs have to be developed and optimized. It seems advantageous to avoid a situation in which no costs accrue to the end user and there is no incentive to be frugal or cautious in requesting materials that may not be essential. Other materials are not renewable, such as some clinical samples, unamplified libraries, extraterrestrial samples, deep sea or deep drilling cores, and fossils. How these samples are treated for possible sharing will have to be dealt with largely on a case-by-case basis. The overall guiding principle in such decisions should be scientific merit and the acquisition of information of interest to the scientific community at large.

Equipment, unlike data and most samples, is saturable. In addition to an incremental use cost, the total amount of available access is limited. Some animal resources are also saturable. For example, the number of animals that a primate center can produce and maintain is certainly finite. Here, a proper balance needs to be struck between acknowledging or rewarding those who had the foresight to construct, acquire, or fund such equipment and the desire to see equipment (or animals) be available for use by the highest-quality scientific projects, wherever they arise. Some facilities, such as synchrotrons, are best viewed on a worldwide basis. Others will be best managed on a national, regional, local, or institutional basis. A general guiding rule that seems applicable to most cases is dividing available time so that those who are responsible for the resource have significant privileged access, but the remaining access to the resource should be competitively available to all users. External use should be judged by scientific quality and by the need for access to the unique resources. Under ordinary circumstances, whether or not an external project is directly competitive with one already ongoing at or planned for the shared facility should not be considered in making this assessment. Occasionally, when a project is extraordinarily taxing in terms of the time or staff available at a resource, competitive projects should be discouraged. Here, the potential competitors should, if at all possible, be encouraged to work as a team. If this fails, first come first served seems like the only simple system to resolve the conflict.

Incentives and Rewards for Resource Sharing Are Not Fully Developed

The current systems for rewarding academicians or employees in industry do not encourage sharing but rather focus on individual achievements. There are no simple answers to questions such as the following: how much "credit" should an individual receive for providing transgenic animals or research reagents to colleagues, and for what period of time? How should the collaborative contributions of individuals scientists to research projects be evaluated?

Sharing Requires Incentives, Not Disincentives

For academic scientists, incentives are citations or other credit for use of samples made available; another incentive is having the costs of making these samples available covered by the recipient, a third party, or one's grant. Incentives also need to be offered for those who make raw data available over the World Wide Web, since some remote reprocessing of raw data will inevitably be quite valuable. A foreseeable shift in emphasis toward more theoretical or computation biology means that the impact of sharing data that is not normally in public databases must be addressed in a timely fashion. Provisions for sharing data, materials, and equipment should be built into research proposals, and the sharing activities should be included as part of the progress report when grants are being considered for renewal. For all sharing of materials, data, and equipment, there is a temporal threshold after which the individual investigator should be removed from the loop (i.e., although soon after discovery, an investigator might reasonably demand coauthorship from others using his or her resource, after some period only an acknowledgment is appropriate).

The willingness of scientists to participate in the *Arabidopsis* project was enhanced by the scientific credit they received for participating as well as the peer pressure exerted upon those who were less enthusiastic participants.

Likewise, the major incentive to an investigator to contribute animals to TJL is that he or she receives considerable scientific credit and also frees him- or herself from maintaining a colony to supply peers with animals. Similarly, it frees him or her from the attendant issues of shipping, monitoring, advising, et cetera. The disincentives are that it increases the competitiveness of scientific peers and is an expense to the contributor—although this may be charged to grant support or may be supported by the contributor's institution. A different type of disincentive occurred when there was an exponential growth in requests for materials from LLNL, which was inadequately funded to support these requests and received little or no scientific credit for providing these resources.

The Importance of Material and Data Assets Changes Over Time

A key clone at the early stages of an investigation may be worth trading only in an actual scientific collaboration. Later, the clone may be freely available in a public repository or distributed upon request. Finally, the clone may become archaic: it should not be kept or distributed; public repositories should deaccession it.

Technologies and Needs Are Evolving Very Rapidly

Any system put into place must have sufficient flexibility to evolve as well.

New Definitions of "Publication" May Have to Evolve to Keep Pace With the New Electronic Information Systems

It is remarkable that over the past two decades at least a millionfold increase in the power of computer hardware and software has occurred without any significant impact on the way credit is awarded in the university research community for work performed and reported. Should the inclusions of methods, sequences, or other data in readily accessible databases have some relative merit compared to scientific articles and book chapters? Ways of providing credit to institutional shared resources must be found, or support for the scientific mission of these core activities—which benefit many—will be endangered.

Methods for Enforcing Existing Policies on Sharing Are Inadequate

Although some policies already exist mandating sharing, most notably that of the Public Health Service in regard to products of research with public monies, the enforcement of these policies is inadequate. It is possible that better rewards for sharing will make failure to share sufficiently unattractive that no explicit sanction is necessary, but until that time it seems only logical to discourage noncompliance at the same time as we reward compliance. Should universities be the main point of enforcement? To what extent should government funding agencies take a role in enforcing sharing? How should the willingness to share impact funding? The role of universities and professional organizations in encouraging and facilitating sharing was prominent in workshop presentations and discussions. Actions against scientists who fail to share, however, are rare. To a very small extent, NIH has required sharing or withheld funding (especially for structural data). It was unclear whether NSF has taken the same position.

The policy stated, but also not rigorously enforced, by many journals that a published clone or other renewable sample should be available publicly is a sound policy. The issues yet to be resolved are the actual mechanism of enforcement and how the costs involved should be paid.

There Are Many Private and Public Stakeholders in any Major Resource Sharing Attempt, Often With Conflicting Goals

The boundary between private- and public-sector activities that impact on shared resources is complex and raises issues that will need to be monitored carefully. The National Center for Biotechnology Information (NCBI) provides an interesting example in which, as a compromise, a federally funded public database will make some software publicly available, but the provision of commercial quality supported software is left largely to the private sector. DNA synthesis and DNA sequencing are two other areas in which the needs of the community and the activities of the private sector will have to be balanced. It does appear that economies of scale will dictate that some such activities are better provided as private-sector services as long as actual costs to the users do not inhibit research.

In WRPRC, the ownership of the monkeys is retained by the institution, but use of the animals resides with the scientists after appropriate peer review. When internal review committees for saturable resources such as nonhuman primates or synchrotrons exist, however, concerns about conflict of interest between internal scientists competing for the same resources must be closely monitored.

Resources may often go to those who possess the most money to pay for them or who have the freedom to profit from them. This may place equally or more creative scientists, who are less well off financially at a disadvantage. For example, the costs of some mouse strains at TJL is driven up when for-profit groups "cherry-pick;" for example, they undercut sales of popular mouse strains by TJL by marketing only the most financially viable animals and ignoring less commercially appealing strains. TJL maintains the latter as a service to the research community, using revenues from the former to offset the costs of maintaining the colonies.

The perception that scientific data and research materials (animals, reagents, etc.) have potential commercial value frequently causes universities to be even more reluctant than individual scientists with respect to sharing.

The relationship between intellectual property issues and sharing is a complex one. It rests on ambiguities in current issues of credit and ownership that go beyond the additional constraints imposed by sharing. These issues are badly in need of clarification and resolution. One example is the status of the research exemption from licensing for university-based investigations in a climate where universities are required by law to protect intellectual property that is potentially valuable commercially. The current status of PCR patents is one area ripe for such investigation.

It must also be recognized that different cultures regarding sharing may exist within academia, or industry and among individuals scientists irrespective of their place of employment. Industry is generally thought to focus heavily on retaining intellectual property rights by stringent enforcement of confidentiality and material transfer agreements. Efforts to protect the long-range interests of stockholders may involve demanding far-reaching agreements that make ownership claims on future inventions related to the material or technologies industry produces. The activities of Bristol Myers and the human immunodeficiency virus strain HIV-2287 are an example. Other companies have demonstrated a more thoughtful, long-range concept of value. The government and the scientific community should seek ways to foster this more enlightened attitude.

Who Pays and What Do They Pay for?

The issues of quality control and quality assurance for shared samples or sample repositories are of major concern. Sharing of individual reagents even within single laboratories is often compromised by concerns about improper prior sample handling. Both TJL and ATCC have resolved these problems by characterizing the animals and materials they provide. Mice from TJL are of known genetic background, which is constantly monitored, and have been caesarean rederived to eliminate diseases that will affect research results. JAX mice may be more expensive than those from other suppliers that do not provide the same quality. Commercial competitors willing to employ less stringent measures on a smaller selection of resources can and do offer apparently similar products at cut-rate prices. High-quality research depends on high-quality materials, and the scientific community will have to recognize that it must pay for quality control, through subsidy if not through user fees. Similar issues regarding quality control may exist for shared data. How well are the data validated?

No simple universally applicable answer emerges, but a combination of improved analytical tools for quality assessment and user education about proper sample handling methods will help to reduce costs incurred by wasted or contaminated samples considerably. It is worth noting that for chemicals or reagents, where any kind of hazard is involved, the cost of disposal often dwarfs the cost of acquisition. This argues strongly for virtual supplies, stockrooms, or repositories where samples are not created or subdivided until they are needed. Such a scheme will work only with an extremely efficient distribution system. The use of electronic ordering, inventorying, and purchasing will become the norm, and this should help encourage efficiency.

A key ingredient to quality control is the funding for key administrative and support personnel who carry out this essential, but relatively low-profile, activity.

Regulatory Requirements and Documentation Can Be Unnecessarily Complex and Burdensome

Regulations promulgated by government agencies affect shared resources disproportionately. The regulatory burden on ATCC for shipping many samples is necessarily greater than that on an individual who ships an occasional sample. Some regulations governing animal care and shipping by the various municipal, state, and federal agencies are conflicting. Regardless of their scientific basis, the costs of complying with these regulations and the extra documentation required by them add burdens to the individual scientist, his or her institution, and the shared resource. Among the underlying reasons for the centralization of primate center programs, for example, was the desire to increase animal welfare and decrease cumulative regulatory costs; despite this, regulations and requirements for documentation for the use of animals or animal tissues continue to increase exponentially.

A second issue, only tangentially addressed in the workshop but potentially stifling to some sorts of clinical research, is the increasing regulatory activity regarding the use of human tissue and tissue products in research. Who "owns" these materials and what sorts of informed consent must be obtained before they are used or reused?

Education of Scientists Covers Neither the Ethos of Sharing Nor Intellectual and Tangible Property Issues

There is a significant gap in leadership in the training of scientists with regard to the issues of intellectual and tangible property: What constitutes intellectual property? When and how can (or should) patents be used to protect individuals and institutions? During training, there is no formal emphasis on the merits of sharing or the benefits of collaborations, and in an increasingly competitive atmosphere where resources are limited, the benefits of sharing may be unappreciated.

Resource Sharing Can Have National and International Implications

What are the consequences on the U.S. position in international trade of complete government funding of national culture collections (e.g., in Germany and Japan)? What guarantees are there of future access by U.S. companies, and individual scientists? In various countries the relationships between business and government differ, and the support for core shared facilities that benefit business often derives from the government. How will such national authorities interact with countries such as the United States that are in turn providing resources to them? What benefit is there to various governments to duplicating databases and collections? How will countries that have different interpretations of intellectual property treat scientists from other countries? What protection can these scientists anticipate? In underdeveloped countries will the desire to protect what are perceived as national resources, such as plants or animals, impede the free movement of materials and animals?

Wherever resources are saturable or irreplaceable, all efforts should concentrate on viewing the scientific utility of such resources from a worldwide perspective. Procedures should be developed for worldwide review of competing applications for limited resources or facilities. Synchrotron x-ray sources are one area ready for the early implementation and evaluation of such procedures. Ecological and environmental samples, and strain collections are other areas in which a worldwide perspective is absolutely essential. The United States is in a strong position to catalyze such global efforts because, today, it has a major position in shared scientific data—a valuable resource that is already made available on a worldwide basis. The National Research

Council is in an excellent position to work to realize these goals by networking with other academies and relevant government agencies worldwide.

There Is a Gap in Leadership

Sharing of research resources lacks high-profile leadership (for example, the president of a major scientific society or the president of the National Academy of Sciences). Universities, government agencies, and industry have failed to focus the scientific community.

Partnerships in Sharing Resources May Be Unequal

The issue of fairness in access and opportunities to utilize shared resources is ongoing because there are typically inequities between those seeking access to saturable or costly resources. For example, graduate students or junior faculty may seek resources from large companies or senior investigators but have little to offer by way of a collaboration, whereas a more senior investigator seeking the identical resource may be perceived as an attractive collaborator.

Monopolies Can Be Good or Bad

Federal funding policies typically require competition for funds, but in some cases this may be an artifice that is unwarranted. Although a competitive renewal of primate centers might elicit some creative new ideas, it seems less certain that requiring individuals to submit proposals that will compete in setting up stock centers and services for *Arabidopsis* is serving either science or taxpayers well. The goal should be to identify the most cost-effective methods and highest quality resources for the scientific community.

RECOMMENDATIONS

This study is exploratory in nature rather than definitive. The committee was not asked to provide solutions so much as identify present and future obstacles and point out directions for followup in more definite studies by a similar committee or others. The committee believes the Academies are in a unique position to provide leadership and bear some responsibility for the culture and ethos of sharing. As a result the committee recommends study or further work to address a number of the problems and issues raised in the workshop summarized in the previous section.

Administrators of research institutions, grant administrators, scientists, and industry representatives should meet to develop policies to foster sharing of resources. These policies should explicitly address the following:

- **Sources of reliable funding for provision of materials and services to the research community.**

A portion of the costs of sharing should certainly be borne by the requestor of the material or service. In some cases such user fees might cover the entire expense incurred by the provider, but in other cases setting fees at that level would effectively preclude sharing with much of the nonprofit research community. Several of the case studies instead subsidize the sharing of materials, equipment, or services from funds the primary purpose of which is not sharing, just as individual scientists use research grant funds to provide materials to colleagues. Funding agencies should consider more straightforward mechanisms by which facilities might be reimbursed for the full costs of sharing with the rest of the scientific community. One possibility might be peer-reviewed distribution contracts providing reimbursement of costs not covered by user fees. The duration of such contracts should be long enough so that grant writing is not a major activity of the facility, and the need for competitive bidding not so great as to preclude awards to a single competent facility.

- **Training and education regarding the ethos and the value of sharing and related intellectual property issues, including the merit of patents and licensing**

Education in these matters needs to begin early in graduate training and should parallel educational offerings in the area of scientific integrity. As with scientific integrity, education in scientific sharing needs to be strongly reinforced by an environment within the institution that demonstrates willingness to share and the benefits to be derived from such behavior. Ergo, university administrators as well as scientists need this education and training.

- **Rewards and incentives for researchers who share resources**

To foster an environment that can serve as a model for the appropriate education of graduate students and induce researchers to share, it is necessary to develop incentives for those who do share. This means that there must be recognition in terms of academic credit, promotion, and salary for those who share. As a concrete example, acknowledgment for having provided a critical reagent in a significant paper should carry a proportional benefit relative to having been an author of such a paper. In the same manner, funding agencies could make resources available on the basis of such sharing, perhaps by requiring applicants' biographical sketches to include such items as provision of resources to other scientists or repositories and memberships on shared resource steering committees. Deans, department chairs, and other university administrators might then come to view membership on such committees as a prestigious appointment similar to membership in a study section. Grants might also provide additional funds to cover expenses incurred in sharing materials with other scientists.

- **Mechanisms for enforcing agreed-upon resource sharing policies within and across institutions**

The funding agencies have a clear stake in promoting the optimal use of research resources, and in some cases already have articulated clear policies mandating sharing. They are however ill-equipped to investigate allegations of violations, and have as a penalty for noncompliance only the all-or-none revocation of funding. Because the local research institution controls the employment, reimbursement, academic rank, and space available to the researcher, it is potentially the most effective enforcer and in the best position to determine the extent of enforcement required. Research institutions, however, as well as the scientific societies and journals that provide scientists with recognition, do not have the same obvious stake in sharing as the funding agencies. The funding agencies may therefore have to begin this task by arranging a stake in sharing for these institutions. The resulting cooperation would have a synergistic effect regardless of the extent to which both institutions and funding agencies should encourage or insist on sharing.

- **Role of the technology transfer office in facilitating resource sharing**

In several instances during the workshop the statement was made that the institutional technology transfer office was often more of a hindrance to sharing than the individual investigator. Clearly, the technology transfer office has the obligation of protecting the researcher and the institution with regard to intellectual and tangible property; however, there has already been significant progress in the development of uniform material transfer agreements between not-for-profit institutions. This and other such mechanisms can foster sharing and should be aggressively developed and used. Similarly, the strong "advertising" programs of many of the case studies suggests an important proactive role for technology transfer offices, publicizing resources at their institutions available not simply to for-profit partners but to scientists at other academic institutions as well.

- **Current National Institutes of Health guidelines governing university-industry relationships**

The current NIH guidelines governing relationships between universities and industry encourage institutional patenting of NIH-sponsored research results and licensing to industry. Thus, the question arises of the extent to which reagents and results originally dependent on public support should be shared versus the initial period of confidentiality sometimes required for the effective technology transfer intended by current federal regulation.

Federal and private funding agencies and industry should jointly undertake a suitable cost-benefit analysis and explore mechanisms to enhance the efficiency both of funding shared resources and of sharing resources.

A major argument for the sharing of resources is the enhancement of both the effectiveness and the efficiency of doing research. To justify funding of resource sharing, it is necessary to be able to document the savings achieved. The capital investment needed and the demand for the product will help determine the number of and placement of facilities. For instance, synchrotrons by their very nature will be limited in number, and the same is likely to be true for primate centers. Culture collections may offer economy of scale, which would serve to limit their numbers.

Because of the growth of economic nationalism and to avoid unnecessary duplication, the world scientific academies should convene to identify barriers to sharing resources across national boundaries and should develop mechanisms to overcome them.

Ideally science is international. Historically, barriers to exchange of ideas, results, and reagents have resulted from concerns of national security. More recently, economic security has become a more prominent component of national security, and science has come to be appreciated as a major contributor to economic well-being. Appropriate user or sample fees and ground rules for partnerships between industrialized and developing countries demand attention. New culture collections being established in Germany and Japan will be totally funded by the government, raising concerns both about unnecessary duplication and about the possibility of restrictions on the sharing of reagents in the future. Also of concern is the establishment of universal rules for the protection of intellectual property and a commitment to adhering to such rules. The overall issue demands rapid action on the part of the scientific community to forestall decisions at the national level that may be difficult to reverse.

Because the private sector will continue to have a major impact on resource sharing, representatives from industry, nonprofit institutions, and funding agencies should be brought together to work toward solutions of current problems such as the following:

- **Overreaching claims on future ownership of inventions by providers of shared resources and research tools**

The major question is at what point the original provider no longer has a legitimate claim. This includes issues of how far reaching the licensing rights of the provider are and how long the sharing of a resource entitles participation as a full collaborator.

- **Competition between private-sector activities and public shared resources**

At what point should the distribution of a scientific resource be done by the private sector. Currently there is concern about "cherry-picking"—allowing public resources to do the hard work of development and quality control, only to have private businesses undercut these costs by taking advantage of the work done by the public resources.

- **How to protect the research exemption for licensed intellectual and tangible properties**

To what extent should there be a distinction between the use of resources for nonprofit research as opposed to work done for commercial development? If such a distinction should be made, how can that be achieved and what should it entail?

Impediments to biomedical research and education caused by confidentiality requirements

Have confidentiality requirements actually impeded research? Have they done damage to collegiality? To what extent is the lack of sharing caused by commercial concerns versus a more general unwillingness to share?

A cost-benefit analysis should be conducted to evaluate the possible impediments to resource sharing caused by government regulations.

The major considerations should be the extent to which such regulation actually contributes to the desired end, whether the desired end could be achieved in a more economical manner, and finally, whether the benefits really are commensurate with the costs.

Appendix A

Workshop on Resource Sharing in Biomedical Research

National Academy of Sciences
2101 Constitution Avenue, N.W., Washington, D.C.

AGENDA

DAY ONE: JANUARY 22, 1996

8:30 a.m.	Welcome and Opening Remarks on Resource Sharing **Bruce Alberts, Ph.D.** *President, National Academy of Sciences*
9:00	Resource Sharing: With Whom, When and How Much? **David Cordray, Ph.D.** *Professor of Public Policy and Psychology* *Vanderbilt University*
9:45	Case Study #1 American Type Culture Collection **Raymond Cypess, Ph.D.** *Chief Executive Officer*
10:45	Case Study #2 Multinational Coordinated *Arabidopsis Thaliana* Genome Research Project **Chris Somerville, Ph.D.** *Carnegie Institute*
11:30	Case Study #3 The Jackson Laboratory Animal and Genetic Resources **Muriel T. Davisson, Ph.D.** *Senior Staff Scientist and* *Director of Genetic Resources*

1:00	Panel on the Role of Journals in Promoting Sharing ***Herbert Tabor, Ph.D.*** ***Editor, Journal of Biological Chemistry*** ***Jerome Kassirer, M.D.*** ***Editor, New England Journal of Medicine***
1:45	Case Study #4 Regional Primate Research Center at the University of Washington-Seattle ***William R. Morton, D.V.M.*** ***Director***
2:45	Case Study #5 Macromolecular Diffraction Biotechnology Resource Cornell High-Energy Synchrotron Source (MacCHESS) ***Steven E. Ealick, Ph.D.*** ***Principal Investigator***
3:30	Case Study #6 Human Genome Center at Lawrence Livermore National Laboratory ***Anthony V. Carrano, Ph.D.*** ***Director***
4:15	A Role for the Private Sector in Sharing Scarce Resources? ***David Barry, M.D.*** ***Chief Executive Officer*** ***Triangle Pharmaceuticals, Inc.***
5:00	An Electronic Clearinghouse for Research Materials Exchange ***Eugene Sokourenko, M.D., Ph.D.*** ***President*** ***LabSearch International***
6:15	Dinner The Role of Government in Promoting Sharing ***Harold Varmus, M.D.*** ***Director, National Institutes of Health***

DAY TWO: TUESDAY, JANUARY 23

8:30 a.m. Guided Discussion of Key Topics in Light of Case Studies
(See attachment for explication of key topics)

       Ownership—Guides:
          Russell Ross, Mark Frankel
       Access—Guides:
          Queta Bond, Allan Shipp
       Function—Guides:
          Ken Berns, Judy Vaitukaitis
       Costs and Cost Savings—Guides:
          James Knighton, David Martin, Francis Meyer
       Future Starts and Stops—Guides:
          Charles Cantor, Marvin Snyder

12:15 p.m. Workshop Adjourned

1:15 **EXECUTIVE SESSION:**
Committee meets to discuss outline of
final report and make writing assignments

3:30 Adjourned

Appendix B

Acronyms

AAALAC	American Association for Accreditation of Laboratory Animal Care
ABRC	*Arabidopsis* Biological Resource Center
AIMS	*Arabidopsis* Information Management Center
APAD	Acquisition, preservation, authentication, and distribution
ATCC	American Type Culture Collection
AtDB	*Arabidopsis thaliana* Database
BSL-3	biological safety level 3
CCD	charge-coupled device
CESR	Cornell Electron-Positron Storage Ring
CHESS	see MacCHESS
CRS	Collection, Research, and Services
DOE	U.S. Department of Energy
EST(s)	expressed sequence tag(s)
FISH	fluorescence in situ hybridization
HIV	human immunodeficiency virus
IACUC	Institutional Animal Care and Use Committee
IDA(s)	international depository authority(ies)

IMAGE	Integrated Molecular Analysis of Genome Expression Consortium
IMR	Induced Mutant Resource
IOM	Institute of Medicine
ISDB	integrated scientific database
JAX	Jackson Laboratory mouse code
LLNL	Lawrence Livermore National Laboratory
MacCHESS	Macromolecular Crystallography Resource at the Cornell High-Energy Synchrotron Source
MAD	multiple wavelength anomolous diffraction
MIR	multiple isomorphous replacement
MSDN	Microbial Strain Data Network
NAS	National Academy of Sciences
NASC	Nottingham *Arabidopsis* Stock Centre
NCBI	National Center for Biotechnology Information
NCRR	National Center for Research Resources
NIH	National Institutes of Health
NMR	nuclear magnetic resonance
NRC	National Research Council
NSF	National Science Foundation
OPRR	Office for Protection Against Research Risks
PCR	polymerase chain reaction
R&D	research and development
STS	sequence tagged sites
TJL	The Jackson Laboratory
USDA	U.S. Department of Agriculture
WDC	World Data Center
WRPRC	Washington Regional Primate Research Center
YAC(s)	yeast artificial chromosome(s)